Voyage to the Grand Banks

The Saga of Captain Arch Thornhill

Raoul Andersen

Voyage to the Grand Banks

The Saga of Captain Arch Thornhill

Raoul Andersen

Creative Publishers
St. John's, Newfoundland
1998

© 1998, Raoul Andersen

The Canada Council | Le Conseil des Arts
для the arts | du Canada
since 1957 | depuis 1957

We acknowledge the support of the Canada Council for the Arts for our publishing program.

We acknowledge the financial support of the Department of Canadian Heritage for our publishing program.

All rights reserved. No part of this work covered by the copyrights hereon may be reproduced or used in any form or by any means—graphic, electronic or mechanical—without the prior written permission of the publisher. Any requests for photocopying, recording, taping or information storage and retrieval
systems of any part of this book shall be directed in writing to the Canadian Reprography Collective, One Yonge Street, Suite 1900, Toronto, Ontario M5E 1E5.

Cover: "On the Banks of Newfoundland"
Courtesy: Centre for Newfoundland Studies Archives, Coll-137
Memorial University of Newfoundland
Design: David Peckford

∞ Printed on acid-free paper

Published by
CREATIVE BOOK PUBLISHING
a division of 10366 Newfoundland Limited
a Robinson-Blackmore Printing & Publishing associated company
P.O. Box 8660, St. John's, Newfoundland A1B 3T7

First printing December 1998
Second printing June 1999

Printed in Canada by:
ROBINSON-BLACKMORE PRINTING & PUBLISHING

Canadian Cataloguing in Publication Data

Andersen, Raoul, 1936-

    Voyage to the Grand Banks

    Includes material from interviews with Cpt. Thornhill
    Includes bibliograpshical references and index.
    ISBN 1-895387-25-6

1. Thornhill, Archibald, 1901-1976.
2. Fishers — Grand Banks of Newfoundland — Biography.
I. Thornhill, Archibald, 1901-1976. II. Title.

SH224.N7A52 1998   639.2'2'092   C99-950004-X

*To those whose lives
build our heritage
of the sea.*

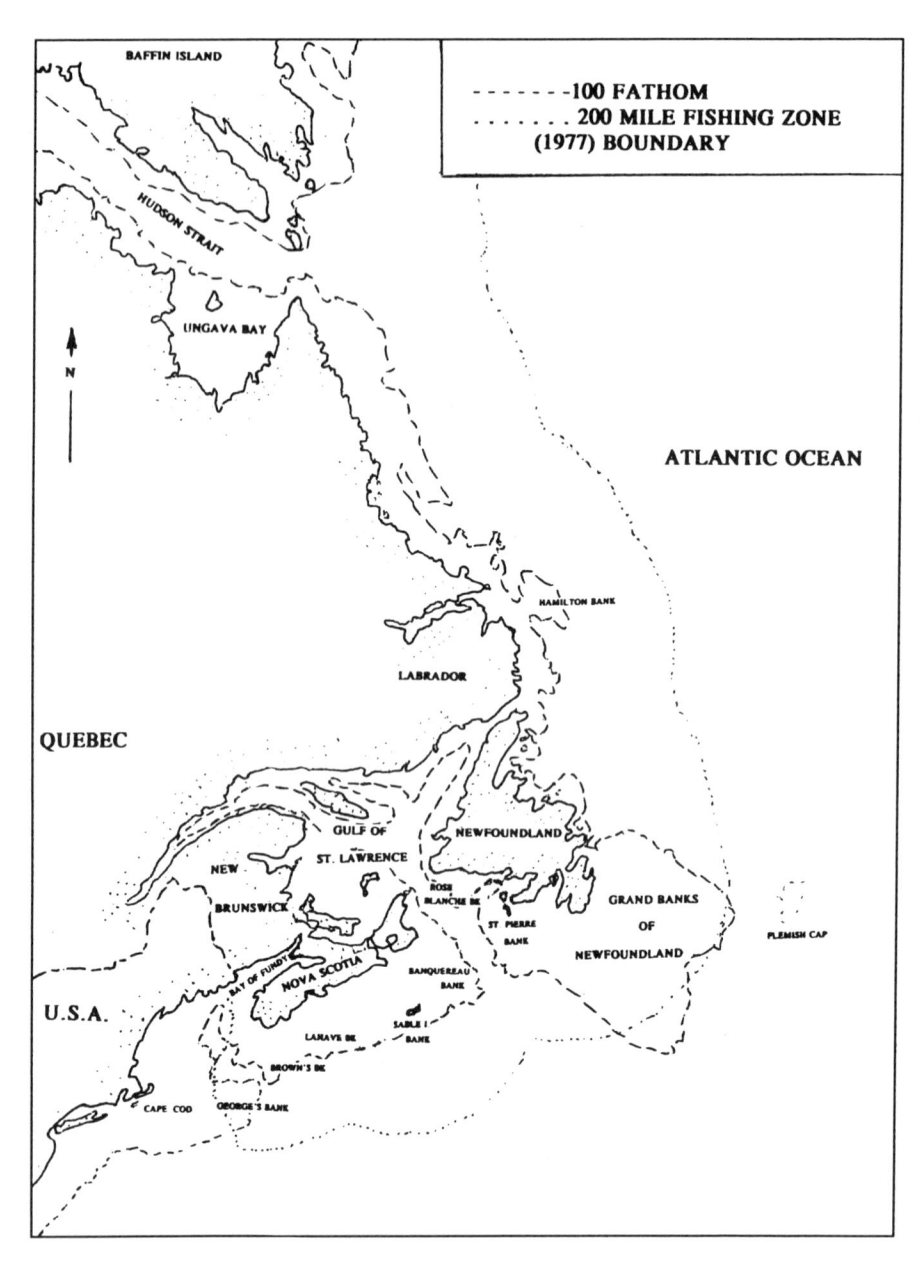

NORTHWEST ATLANTIC FISHING GROUNDS

CONTENTS

Preface ix

Introduction 1

1   An Excellent Place for Fishing 15

2   I Never Gave Up Once in My Life 35

3   Voyage to the Grand Banks 67

4   My First Command 103

5   Barbados She's Goin'? 129

6   It's Just a Six Hour Journey 153

7   Never Lost a Line, Never Lost a Hook . . . 167

8   Hang On a Little While 179

9   Another Dummy Amongst the Big Power Vessels . 191

10 A Wonder for Her Time 211

11 Millions of Fish 241

12 The Voyage Ended 281

Endnotes 295

References 327

Appendix 330

Index 333

Glossary 347

PREFACE

This book resulted from a collaboration between an outport-reared Newfoundland deep sea fisherman, Captain Arch Thornhill (1901-1976), and a 'come from away', city-bred social anthropologist. The Captain's career bridged inshore and offshore banks fishing under sail, then groundfish trawling under power. His career brought him to the leading edge of major changes in the industry. He intended this account we were developing together as a testimony about his generation's experiences in Newfoundland's changing twentieth century deep sea fishery. Following the Second World War access to growing fresh fish markets moved this industry from underdevelopment to the limits of its fish resources in what became an explosive and uncontrolled industrial expansion.

In 1967 I began study of Newfoundland's offshore trawler fishery to learn about the organization of its industry and how its fishermen adapted to what was often depicted as one of the most hazardous industrial occupations. At the time there was little social science literature on deep sea fishermen. The leading contemporary account, Jeremy Tunstall's, *The fishermen* (London: MacGibbon & Kee, 1962), described men in the distant water trawler fishery carried on from Hull and Grimsby, England, chiefly for fresh fish markets. When I first read it in the early 1960s I was already interested in deep sea fishing and in Newfoundland's fisheries. I wondered how what Tunstall described of this English industrial occupation fit Newfoundland's trawler fishery?

In the event, Newfoundland's trawler fishermen employed similar fishing technology in a different operational regime, and their social and cultural identity was rooted in unindustrialised small coastal communities only recently freed from an economy and society where fish merchants and banks schooner owners had exercised control by giving out supplies on credit in exchange for fish, chiefly cod, most of which was salted and dried for export to distant markets. Before the advent of trawlers the work of Newfoundland's deep sea fishermen — fishing from dories and their schooner motherships —compared yet differed from that of their Nova Scotian and New England contemporaries. But it was every bit as hazardous. As Skipper Arch observed, "It was a hard life, my son. But, still, there was something about it."

By mid-century the chemistry of life and labour in Newfoundland's outports had been altered by, among other things, the rise of a cash economy, confederation with the Canadian industrial state in 1949, rapid adoption of mechanized groundfish trawlers and a developing fresh-fish industry that promised year-round employment. Newfoundland's deep sea fishermen were well along toward fully industrial work, community, and individual lifestyles.

However, the way they viewed their occupation and the new industry at this time was informed by the banks schooner and dory fishery and small community lifestyles they had left behind only recently. Modern trawling, by contrast, was carried on from what government called 'growth centres'. This fisherman's saga adds to our understanding of the industry, work, conditions, culture, and men in these two fishing eras, and the relations between them in the transition to modern trawling.

Captain Thornhill's story also speaks to the major fish resource, industry, and cultural crisis of our time: the destruction of eastern Canadian groundfish stocks. It helps illuminate forces that lay behind the collapse of Newfoundland's fish resource base.

Fishing and culture ways in transition and collision

As the Introduction below indicates, specific work on Skipper Arch's story began in 1975, shortly before Canada decided it would create a 200 mile exclusive fisheries management zone. Eastern Canadian offshore groundfish stocks then were in peril. The International Commission for Northwest Atlantic Fisheries, ICNAF, had failed to control the offshore harvest by both domestic and foreign groundfish trawlers and factory ships, there were concerns inshore and offshore about intercepted fish and stock declines, and a growing international sentiment that states adjacent to fish stocks and whose people are most dependent upon them might best be charged with their management.

In 1977 Canada embarked upon management under the 200 mile zone with much optimism. The principle of coastal state preferential fishing rights would enable and require Canada to expand its trawler fishing fleet and processing capacity, foreign trawlers would be phased out of the new fishing zone as Canada's harvesting capacity expanded, its world class fisheries science and new enforcement opportunity would turn fish stock decline around, and potential Canadian landings would be so great within a decade or so that its producers would have difficulty finding markets.

Against this background of anticipated opportunity Newfoundland's fishing outports and industry — from the inshore to offshore sectors — could look to better times. Most Newfoundland outports at this time pursued seasonal inshore fishing in an atmosphere of disillusionment about the future of the industry's inshore sector, its workers and communities. Now the old "burn the boats" and leave the fishery attitude of earlier decades was suddenly turned on its head. The Honorable Romeo LeBlanc, then federal fisheries minister, had struck a public stance favourable to developing the inshore fishing sector while checking the destructive tendencies of big trawler companies, or so it seemed.

A mere twenty five years later, however, in 1992 Canadian federal fisheries authorities were compelled to admit their management efforts had failed, its Maritime provinces, and Newfoundland fishermen faced a marine ecological disaster, and they declared moratoria on its major commercial fisheries. There were too many boats and fish processing plants and too few fish, and their stock biomass was in perilous condition. There was talk of "commercial extinction" with unknown consequences.

Overnight, the livelihoods of tens of thousands of Newfoundland and other Canadian workers were suddenly yanked away. It has been an industrial employment disaster unprecedented in Canadian history, and all the more bewildering because it happened under federal government management. And many suspect it was intentional, designed to get fishermen out of boats, and effect resettlement of Newfoundland's scattered outport fishing population by destroying their resource base. Thus the *sine qua non* of Newfoundland's way of life, its culture, was being destroyed.

Efforts to understand what went wrong, to help those affected regain an economic footing and adjust to this massive human tragedy, to restore annihilated fish stocks, and to chart a new course for the fishery's future continue at this moment amid great uncertainty. Along the way, the offshore trawler fishery, once the vanguard of future prosperity, became a villain. And both the ecological disaster and ameliorative — "industrial adjustment" — programme measures fuel a small industry producing literature and proclamations about its issues.

The serious human failure to manage our renewable resources is underlined by the knowledge that fishing people around the globe face similar fisheries crises. The illusion of superior science, technology, and fisheries management in advanced industrial states has been dashed. We haven't been very smart with our common property. What to do? Who will teach us? Who to trust? What fits and what makes sense? Will we learn? Who should or will decide? Gratuitous suggestions from afar are easy. As this is written Canadian authorities have announced plans to decentralise their fisheries management to enhance provincial input and create a more responsive structure. The idea evokes changes in the 1970s with similar aims. What will the outcome be?

Whatever become our answers and decisions, their major burden will fall upon people in the fishing industry and its dependent communities. Whatever we make of early fishermen views that stocks were in danger long before government scientists and managers realized it, their voice must figure more effectively in shaping the future.

About the text

Biographies and autobiographies are windows onto other lives and times. By standing in their subject's shoes, we may see a society and culture as the individual saw, navigated and helped create it. Whether written or oral tradition, they are both history and performance, and always incomplete. Yet they seem essential to ordinary lives. They contribute to mutual- and self-understanding, a sense of identity, and living in our time.

Life-telling ventures begin and end in different ways. This is one of them. The present story of Captain Thornhill's life and career weds information from tape-recorded and unrecorded oral interview-discussions with Captain Thornhill and others who knew him, directly or indirectly, and the changing fishing industry of which he was a part; the captain's written recollections and notes; and published and unpublished documentary sources. The manuscript for this story was completed largely after the captain's death in 1976. It was mutually understood that I would edit the original oral and written texts of this developing story for clarity, coherence, and historical accuracy where possible. Dialect preservation was not a priority. Its introductory commentaries, final organization, and addition of ancillary information are my responsibility.

My work with Captain Thornhill began with several exploratory interviews at his St. John's' home, on June 24, September 30, and November 10, 1975. Although I came armed with preliminary general and specific questions, I used an open-ended approach, always seeking new questions. I took written notes and tape recorded as much as possible.

At the start neither the Captain nor I were certain of the outcome of these initial interviews. We shared the problem of developing confidence in a worthwhile undertaking. I had to commit myself to what might become a long-term project interrupted by immediate teaching and other research responsibilities.

The issue was different for Captain Thornhill, who had to be convinced that a detailed written account of his experience in the Newfoundland fishery had some useful purpose. Fortunately, we shared the view that the present younger generation of Newfoundlanders did not comprehend what his generation had to contend with in order to make their living. The absence of a clear record of Newfoundland's banking schooner and dory fishing operations made it difficult to understand the older generation's views of the present and its problems. I believe the Captain accepted the argument that rendering an account of his life as representative of his generation might help bridge this gap in inter-generational understanding.

During our 1975 meetings I formed a rough chronological overview of Captain Thornhill's career to guide subsequent interviewing. It also seemed wise to create a test development of information from our meetings, so I constructed working drafts of what might become chapters for a full account of his life. The technical content of much of our exchange compelled me to prepare transcriptions ready for editing and working drafts of what eventually became chapters about: Navigating the *R.L. Borden* to Barbadoes in 1933 (see Chapter 5); his unsuccessful struggle to sail the *J.E. Conrad* from Grand Bank to Burin in November 1939 (see Chapter 6); and facing an August hurricane in the *Florence*, and being dismasted and losing her in 1939 (see Chapter 8). Chapter 8 is perhaps the most refined reconstruction of events in this volume. With early publication in mind, we revised several drafts and the galleys for accuracy and coherence.

These test chapters inspired Captain Thornhill to take up pencil and scribbler. He reconstructed other sections of his past to "fill in" the story, beginning with his earliest years in Anderson's Cove, Fortune Bay, to his years aboard the dragger *Blue Foam*. These additions are the substance of Chapters 1 to 4, 7, and 9 to 11. I felt our work had become a true partnership.

Like other Newfoundland banks skippers, Captain Thornhill had kept no detailed log or diary during his working years. I was asking him to reconstruct the past from memory and he didn't want to be faulted for inaccuracy. Responding to this challenge, Ruth, his wife, quietly contributed much from her own recollection of events along the way. I undertook to document events described and incorporate information from our interviews into the developing text to elaborate and clarify selected points. But our purpose is to add Captain Thornhill's authoritative voice to a growing discourse about Newfoundland's fisheries and people. His story raises many issues that invite further historical and/or social anthropological examination in their own right. Ancillary information given in Endnotes touch on some of these issues. But further analytic discussion would require a very different book.

Captain Thornhill's sense of propriety dictated that what we were writing not embarrass his family or other individuals. For example, it might be interesting to learn *who* did or said this or that, as with the hard drinking "laddio" banks skipper mentioned in Chapter 3, but it was unnecessary to our purpose. Likewise, over forty years after occasions when dory crew members ended their voyage with the rank of "low dory," Skipper Arch said, "I can name the names as if it was yesterday, but I'm not going to. I'm not mentioning any names". With some exceptions, such information has been omitted.

In winter 1976, I flew to his winter home at Gander, for further discussions, and our last session together occurred on June 9, 1976, at his home in St. John's. Skipper Arch was recovering from the flu, and our talk on this occasion was brief. He was then working on a draft covering his years aboard the *Blue Foam*, and looking to a summer holiday at the cabin of his son, Roland, outside Darmouth, Nova Scotia. His old and dear friend and brother-in-law, Captain Clarence Williams, the man who taught him navigation just prior to his trip to Barbadoes in 1933 (Chapter 5), had suddenly died a few weeks before and he felt his loss most deeply. Early on the morning of June 29, Captain Thornhill had a fatal heart attack.

Following his death the scribblers containing the handwritten account of his first *Blue Foam* years (1948-51), and other documents about his career, were given me by his wife, Ruth, for use in completing this book as Arch and I had planned.

Parts of Captain Thornhill's career account are given in three earlier publications. They are: "Bound for Burin." *Newfoundland Quarterly* (1977), Vol. 73 (4):17-22; "Millions of fish" (the transition to modern fishing in the Northwest Atlantic, a fisherman's account), *Newfoundland Quarterly* (1980), Vol, 76 (2):17-24. (Reprinted in *Canadian Issues (1980)*, special issue on "Canada and the Sea," vol. 3, No. 1 (Spring): 127-139); and "'Chance' and Contract: Lessons From a Newfoundland Banks Fisherman's Anecdote." In *Merchant credit and labour strategies in the staple economies of North America*. R. Ommer and C. Vickers, editors, pp. 167-182. Fredericton, N.B.: Acadiensis Press, 1990.

"Bound for Burin" is the substance of Chapter 6 in this present volume. Constructed solely from interviews, it recalls an especially telling anecdote from Captain Thornhill's life. It had been prepared and accepted for publication in the *Newfoundland Quarterly* during the fall 1975. Early news that it would appear in this popular Newfoundland historical periodical encouraged Captain Thornhill about our project and to take up pencil and paper himself.

"'Chance' and 'contract'"(1990) is about the first agreement struck by Arch in 1918 between himself as banks dory fisherman and his dory skipper, and the Grand Bank firm that owned their schooner. It embraces parts of Chapters 2 and 3 in this volume. It too began as an oral account to which the Captain later added in writing. "Millions of fish" (1980) tells of Arch's first trawler command in 1948. It combines the unfinished contents of Arch's written account prepared in May 1976 with information from published sources and interviews. This is Chapter 11.[1]

The long term value of life accounts

Apart from its popular and academic value, there was an unexpected bonus to information collected from the late 1960s into the 1980s, from Captain Thornhill and other individuals familiar with Newfoundland's south coast fishery early in this century. For example, in 1989-90 this information suddenly had special value as background information for Canada's negotiating position in the Canada-France Maritime Boundary Arbitration. I hope Captain Thornhill's saga will continue to be useful to others who wish to know the sea, fisheries and seafarers.

Acknowledgements

This volume was developed over a long time, in fits and starts. I am indebted to the many individuals who contributed to this effort. If you are not acknowledged here, the oversight was unintentional.

My introduction to Newfoundland's fisheries began in 1967-68 while on a Postdoctoral Fellowship from the Institute for Social and Economic Research, at the Memorial University of Newfoundland. I am indebted to Robert Paine and Leslie Harris for this early fellowship support. Other occasional small grants from Memorial University have supported my continued involvement in fisheries-related research over the years.

The work begun in 1975 toward *Voyage to the Grand Banks* was an unfunded research initiative. Following Captain Thornhill's death in 1976, a small Canada Council Explorations award in 1977 (to develop a fisherman's life account) helped me gather supplementary information specific to his fishing career. I made research trips to: Gloucester, Massachusetts, for information about the *Gudrun*, whose loss with a full crew in 1951 is discussed in Chapter 11; Lunenburg, Nova Scotia, for details about its banks dory schooner fishery; and St. Pierre, for information on its use by Newfoundland fishermen. During the fall 1977, I interviewed many veteran fishermen and others about the history of the banks and trawler fisheries operated from Grand Bank and other Fortune Bay ports. An SSHRCC Leave Fellowship in 1981-82 provided time for close examination of archival (trip sheet) records from Job Brothers first trawler operations out of St. John's. These records are located at the Department of Fisheries and Oceans, St. John's. I am indebted to Nauss Cluett, who brought their existence to my attention, and located and provided access to them, and Edward "Sandy" Sandeman for assisting with their interpretation. The Leave Fellowship also made it possible to interview many veteran south coast banks schooner fishermen and others knowledgable about this fishery especially in outports west of Fortune Bay along the Connaigre Peninsula and east to Placentia Bay. I hope to complete a volume based upon this larger body of information in future.

My search for information on the history of dory banking schooner operations from Newfoundland, Nova Scotia, and New England involved correspondence with archives as far south as Newport News, Virginia, and personal visits to archives in Ottawa, Halifax, Troense (Denmark), and, of course, St. John's. These archival inquiries revealed a surprising lack of serious interest in eastern Canadian and New England fisheries despite their historical economic and social importance.

A major exception is the voluminous, yet largely unanalised, collection of papers from Lunenburg's Zwicker firm, held in the Provincial Archives of Nova Scotia, at Halifax. In Newfoundland, when its old south coast banks schooner firms dissolved, their financial records were typically destroyed. Almost accidentally, however, while visiting Eric Tibbo, in Grand Bank, I discovered that the ledger records of his family's firm, Forward and Tibbo, Ltd. (1910-1950), were stored in a shop attic. They could be saved. Shortly after, having arranged for their donation to the university's Archives, the late Nancy Grenville, librarian at Memorial University's Centre for Newfoundland Studies, and I drove to Grand Bank, where we filled her small car with ledgers and transported them back to in St. John's. They are now available for study.

I owe special thanks to Captain Thornhill's sons and daughters, Catherine, Roland, Florence, and Cyril for patient cooperation along the way, and Arch's grandson, Paul Thornhill, who was a continuing stimulus during his undergraduate years at university. Sam Tibbo and his mother, Linda, gave me my first lessons about Newfoundland, Grand Bank, and Fortune Bay; Gus Etchegarry, Paul Russell, Harvey Mauger, and Capt. Billy Brushett, aided my access to

Burin Peninsula trawlers, where dragger Captains Freeman Hatch, Gordon Baker, Curt Mitchell, Martin Hanrahan, and Clayton Skinner carried the green university hand to sea where he filled his notes with their lessons; Captain Bobbie Evans shed light on fishing questions; Curt Forsey, retired Grand Bank banks schooner operator and merchant, opened his door for many fireside chats about the old industry; veteran sailmaker Charlie Patten told me of Fortune Bay banks schooners and seafaring; blacksmith Wilson Osborne shared his deep knowledge of Fortune Bay communities, fisheries and personalities, and introduced me to expert veteran banks dory fishermen like Phil Stoodley and Charlie Parsons, and shore fishermen like Phil Riggs and Harry Bradley, and many others; Arch Snook, veteran banks dory fisherman, told me of Brunet and Sagona Islanders; Myril Herridge spoke of the excitement of the first time he was "taken off the turf" to the banks; Harry Keating and Roland Mathews of Burin-area fisheries and work away, in the forest industry, and Nova Scotia, respectively; Annie Bradley and Fred Rogers provided information on community and banks schooner merchant enterprise in Grand Bank; Florrie Hillier told me of the long days and nights when men are away, deep sea fishing, and when they never return; Arch Williams, long-time telegrapher and postmaster at Pool's Cove, told me of communications around Fortune Bay earlier in this century; Al Power, of St. John's, and late banks skipper, Tom Bartlett of Grand Bank, taught me about south coast seafaring, navigation and much more; Betty and Wayne Hollett loaned photographs of Burin-area fishing premises devasted by the 1929 offshore earthquake; Abbie and Edgar Price and Nellie Green, all Grand Bankers, loaned other photographs, as did Lunenburg Captain Owen Creaser; Bertram Riggs provided the cover photo from Memorial University's archival collection of R.E. Holloway photos; Karl Sullivan introduced me to Skipper Arch, and both he and Les Dean recalled Arch's time with the Newfoundland Fisheries Department.

Folklorist Dr. David Taylor shared ideas and my enthusiasm for fishermen's lives and stories; folklorists Marc Ferguson and Cynthia Boyd helped fill gaps in my knowledge of south coast fisheries traditions; Memorial University's Dr. Bill Kirwin helped me understand what this book is and isn't, and the elusiveness of spoken words when written; biologist Don Steele noted questions, left answered; and Memorial University's late Dr. George Story helped in various ways.

Social anthropologists: Tom Nemec generously shared ideas and knowledge; John Cove provided fieldnotes from his dragger fieldwork from Fortune; Louis Chiaramonte, Michael Gaffney, and Geoffrey Stiles added ethnographic flesh to my knowledge of southwest coast Newfoundland and its fisheries; Gordon Inglis and David Macdonald built The Story of Newfoundland's NFFAW union and leadership; Michael Wireman, Michael Orbach, and James Acheson lent ears to parts of the story long ago; Poul Moustgaard guided me to the Troense Museum Archives in Denmark and information on Danish involvement in Newfoundland's saltfish trade; fisherman-anthropologist Kent Martin shared data from interviews with veteran banks fisherman Charlie Hendrick, from Placentia and burning enthusiasm for fishing matters. Ann Douglas and Marilyn Marshall patiently reproduced many drafts of this manuscript for its author. Finally, Don Morgan patiently prodded and steered the author and *Voyage to the Grand Banks* home.

—Raoul Andersen
Memorial University
St. John's, Newfoundland

Introduction

This book presents a fisherman's perspective on deep sea fishing and the way of life developed around it in the first half of the twentieth century. Its story was first given me over twenty years ago by Captain Arch Thornhill — Skipper Arch, one of Newfoundland's premier deep sea fishermen. Born in 1901, like many of his generation, he was drawn to the deep sea or banks fishery from his earliest years in a small Fortune Bay outport on Newfoundland's rugged south coast. In 1918 he made his first voyage as a dory fisherman aboard a schooner bound for the Grand Banks, and he remained active in the deep sea fishery for more than forty years.

To those familiar with eastern Canada's twentieth century fisheries it is perhaps common knowledge that both Newfoundland and Nova Scotia continued the fishing complex of banks dory and schooner operations from the late nineteenth century through World War I, the Great Depression, World War II, and into the mid-twentieth century. However, it is less commonly realized that the social and cultural organization of their respective operations differed. Further, little is known of the connections between the old banks schooner fishery, modern trawling, and the collapse of key groundfish stocks in Newfoundland waters in the 1990s. Skipper Arch's story helps to understand how the banks schooner and dory fishery in this century was organized in both Newfoundland and Nova Scotia, and its relationship to modern trawling. Given mainly in Arch's voice, his story brings us close to how fishermen experienced and organized their lives around these fisheries.

I first learned of Skipper Arch one misty April night in 1967. It was while on my first trip to Newfoundland's offshore fishing banks aboard the *Grand Prince*, a new, 147 foot groundfish stern trawler operating from Grand Bank, on south coast Newfoundland. I was aboard to begin a study of Newfoundland's then expanding and modernizing groundfish trawler industry, how work and lives were organized around it, and its meaning to those the industry sustained. Captain Freeman Hatch was at the helm on the starboard side of the dark wheelhouse, while I braced myself nearby against a teak door frame. As we pitched and rolled along at about four knots across St. Pierre Bank tethered by the heavy trawl she towed, her hardy skipper and I talked about fishing, its work, and the people who made up this deep sea industry and its history.

He told me how steel-hulled groundfish trawlers came to the Newfoundland offshore fishery in the late 1940s, and displaced the banks-dory fishing schooners on which Captain Hatch had first left the turf of a small Fortune Bay outport in the 1930s to sail to the Grand Banks as a teen-aged dory fisherman. Several older hands in his crew had also fished the banks in dories from schooners. As I listened to him and his crew, I learned how groundfish trawling brought a different way of life to Newfoundland's deep sea fishermen. There was a sense of loss in their observations. For example, it seemed to them that sail-rigged, wooden schooners and dories called for a different, more self-reliant, uncomplaining, hardworking, and capable breed of seafarers. And, somehow, unlike the engine-powered, steel hulks, schooners seemed more like living things. Was it a better way of life?, I wondered.

Introduction 3

Witnesses to an earlier industrial era stood at wharf's edge

I had arrived by cab at Grand Bank the afternoon before we sailed. Entering the town center, the cab threaded its way through a flock of sheep being driven up the main street, then dropped me at a boarding house run by a semi-retired couple, Charlie Patten and his wife, where I deposited my bag. After the long drive down from St. John's, especially over the torturous, pot-holed gravel Burin Peninsula "highway," I wanted some exercise. So I walked the few steps from Patten's door to Water Street, to survey the various general stores, shops, and warehouses that backed onto the harbour. The keen, resonating sound of anvil work drew me to the small blacksmith forge operated by Wilson Osborne.

I leaned in over the closed lower half of the wide Dutch door to peer into the dark interior. The glowing forge illuminated Osborne working his anvil while in discussion with several men seated on barrel tops within. In moments he had invited the newcomer in, learned the purpose of my visit and that I was boarding with Patten. He told me he had spent years at the anvil supplying and repairing parts for inshore fishermen and schooners around Fortune Bay, and now draggers engaged in the offshore fishing, and of earlier visits by folklorists from my university. Then he introduced me to Uncle Billy Riggs, a sturdy, quiet and robust older man who had once been in dory aboard various schooners.

Back at Patten's boarding house for dinner that evening, my crusty, straight-talking host spoke of what Grand Bank was like before the early 1950s, when the Bonavista Cold Storage fish processing plant and groundfish trawler fleet brought renewed life to the local economy they now dominated. Mr. Patten had been a sailmaker in his prime years, when Grand Bank gut was sometimes so full of schooners that you could walk across it. The abandonment of sail-powered schooners left little call for his skill. The old grey, clapboard sail loft he had worked in stood vacant steps away from Osborne's forge. That night my thoughts were fired by these compelling local personalities and their stories — about their past.

The following day, as the *Grand Prince* departed Grand Bank we passed a line of once busy, but now largely dormant and empty, saltfish and gear storage and shipping warehouses, merchant shops and small enterprises that had once served the town's various banks schooner firms. Little commercial activity was visible on the long wharf bordering these buildings and the harbour. It seemed as if the weathered, ghostly, and semi-abandoned witnesses to an earlier industrial era stood at wharf's edge, reviewing the present, anticipating its future. Upon entering Fortune Bay, bound for the banks, soon the last signs of the town were church spires, the tops of large homes of once prominent merchant fishing families, then Grand Bank Cape itself.

He's the grandfather of all this

Now at sea several days later, the radio transceiver, always on to receive a hail, suddenly interrupted: "*Makkovik* to *Grand Prince*. *Makkovik* to *Grand Prince*. This is Arch. Are you there Freeman? Over." The caller was Captain Arch Thornhill aboard a small federal government dragger doing exploratory fishing in the bays along Newfoundland's south coast. Unfamiliar with the caller, I asked Captain Hatch who the "old man" was. He replied, "He's the grandfather of all this. He started this draggin' business." Captain Hatch had sailed with Skipper Arch years before aboard the *Blue Foam*, one of the first large side-trawlers in the Newfoundland fishery.

Skipper Arch had commanded both unpowered, and, later, auxiliary, schooners or 'bankers' operating from Newfoundland and Nova Scotia. In 1948 he took command of the *Blue Foam*, one of the first large steel-hulled groundfish trawlers to operate from Newfoundland. He and her crew made record catches in the early 1950s. And he helped train many of the next generation of trawler skippers, and was widely known among Newfoundland deep sea fishermen for his knowledge and skill. Captain Hatch told me that, if I wanted to learn about Newfoundland's old banks fishery and the early trawler years, Skipper Arch was the man to talk to because "he had done it all."

I noted this suggestion for the future. But we were both at sea at the time, pursuing separate agendas. And I was absorbed by the immediate challenge of learning about the current offshore fishery carried on from four south coast ports on the Burin Peninsula, and other Newfoundland ports.

During the following weeks and months the way of life built around the dory banks schooner fishery, and the transition to modern trawling, were frequent issues in my discussions with south coast Newfoundland banks fishermen and others of Captain Hatch's generation and with younger crew members. It seemed that by understanding the organization of life in these two fishing eras, from the comparison and contrasts between them, one learned how the past lived and merged into the present. Without this understanding my description of Newfoundland's modern deep sea fishery and people would result in a partial grasp of the new fishery's meaning to its present participants in respect to its work organization, its conditions, relationships of men to each other, their trawler and the boat and plant owner-operators, to their families and communities ashore, and the viability of the industry itself. Hence, when opportunity permitted, I sought to learn more of the old dory banks schooner fishery. And doing so firmed my early interest in writing this book. It would be a future task, and it would take time. More important, I was unsure how best to approach that task. A practical solution to this issue emerged over the next few years.

From my first anthropological fieldwork in western Canada I had been attracted to the idea of developing a life history narrative study in some context. By the fall 1975, a successful small experiment with this approach while investigating Bermuda's commercial fishery gave me confidence to apply it to the Newfoundland fishery. But its individual subject or subjects remained uncertain. In the event, this present volume relies upon one expert voice, Captain Thornhill, while instructed by many others.

I'm game to try, if you are (Fall 1975)

Skipper Arch and his wife, Ruth, lived together on the second floor of a two flat on LeMarchant Rd., in St. John's. One fall afternoon in 1975 I phoned him to propose we meet to discuss some matters about Newfoundland's old banks schooner fishery. Karl Sullivan, a mutual friend and Arch's former co-worker in the Newfoundland Department of Fisheries, had already told him that I would like to meet him. There was uncertainty in his voice, "Well, I don't know if I can answer your questions, Dr. Andersen, but I'm game to try if you are." When he suggested we meet after dinner that evening I leaped at the chance. We had lived in the same town for over eight years since I learned of him, yet had never met.

That evening my thoughts raced with anxious anticipation as I drove the short distance to his home. What issues should I raise first? How would we get along? Where would our meeting lead? And I wondered at the wisdom and feasibility of my purpose. I would ask Skipper Arch to share his memory of life at sea with me, perhaps to develop his life story. It was to help build a fuller understanding of the offshore fishery as experienced by those who lived the transition from banks schooners and their dories to modern groundfish trawlers.

Upon ringing Skipper Arch's door I was greeted by a blue-eyed, white-haired man in casual dress. He stood about 5'7" tall and weighed little more than 135 lbs. This was the "old man" of legend, the hard-driving and successful banks fishing skipper. At 6'2" I towered over him. He welcomed me and steered me up the flight of stairs to his comfortable, carpeted living room fronting on LeMarchant Road. Arch introduced me to Ruth, a soft-spoken and attractive woman of about the same age. She was a bit shorter than Arch. In quick order she quietly asked me, "Dr. Andersen, will you have some tea?" "Yes, please," said I. Then she retired to the kitchen, and Arch directed me to one of two easy chairs under the front window and we began to talk.

Our talk progressed slowly that first evening, but an overview of his life in Newfoundland's fisheries unfolded in bits and pieces. They included anecdotes rich with information and lessons about personal career turning points and events while aboard specific schooners that could be pinned down by month and year, and other more general information in response to my direct questions.

Arch clarified many matters about banks schooner fishing. Some basic examples that will become comprehensible in the story that follows are: how dory crews were recruited; relations between fishermen and merchants; how the fishing year was framed in "baitings," "trips," and the "voyage;" fishing "under sail" and "to an anchor;" and the difference between fishing on the "count" or "high and low" and on "shares." At this first meeting, after two and a half hours or so my head was spinning with new information that took some hours to untangle and organize before we would meet again.

At the outset I wondered how patient my distinguished, "salt of the earth" fishing skipper would be with a university-trained professor not of his coastal turf or island? In the event, he seemed as eager to learn about me as I was about him. I was grateful to find I passed the sharp "look him in the eye" assessment he was known for and that he tolerated the ignorance exposed by my often naive and perhaps seeming endless questions. We enjoyed each other's company. Over several meetings it was clear that he was the right man for this project. Our collaboration grew rapidly in intensity, productivity and confidence.

She was always there

During our first meeting Ruth had remained in the kitchen near the entry door during our ranging chat, our "men's talk," intervening only long enough to serve us tea and cake. Yet she was alert to our exchange and frequently volunteered helpful details from her kitchen watchpost. This remained her practice in our subsequent meetings until Arch's sudden death the following year. I mention this because I know she was always there for Arch and her family, helping in countless unrecorded ways. (The experience of Newfoundland women in deep sea fisherman households remains poorly documented.[1]) The reader will sense her direct presence only occasionally in Arch's story, yet its fullness owes much to this highly intelligent woman who shared most of his life and yet chose to remain largely in the background in deference and respect for her partner. The testimony of Arch's voice honours them both.

By the summer of 1976 we had overcome our initial mutual uncertainties and formed a strong sense of purpose. We were about to sketch an overview of his final years as the *Blue Foam*'s skipper, and I expected to explore new questions with him about the groundfish trawler fishery. Upon his sudden death, however, I postponed further immediate action to complete the project to rethink its shape. A succession of other projects intervened, some related to Arch's story, others not, until now.

There are things that you would never forget

Memory is always an issue when building an account of someone's life experience, whatever the focus, and whoever the subject. Details are forgotten, and there is always the danger of telling the past the way it should have been. Skipper Arch's story doubtless telescopes many repetitive and mundane events. But it offers details of experience and actions anchored to specific stages and vessels in his career, and events that give classic lessons about, for example, the struggle, hardships, and dangers of fishing and life at sea. Many of these events have powerful emotive connections.

> "There were lives and lives lost; entire schooners' crews, if it was all put together. Men loaded their dories so deep, putting every fish they could in them, and then they'd

swamp. Many times I came aboard and found that men had lost their lives. I'll never forget. There are things that you would never forget."

That forgetting was difficult is evident in the many vessels he recalled as lost or nearly so during his time at sea. But his story is about understanding a way of life, not memorializing for its own sake. It acknowledges the sacrifice and loss of countless individuals, entire crews at times, many of them friends and acquaintances, with devastating impacts on their families and small communities. This expresses the heavy personal reality of seafaring and deep sea fishing lives, and it adds flesh to historical understanding. Likewise, his struggle to gain a berth and to achieve excellence as a banks dory fisherman and skipper help us understand the difficult steps to command, its burdens, and the value imperatives of this hazardous and competitive occupation. "Work? You **had** to work, by God! If you didn't, you wouldn't get any fish. That's all!"

His experience was typical of his generation

Since this volume pivots around one man's "voyage to the Grand Banks" to illuminate an occupation, industry, and changing way of life in the past, the representativeness of Arch's experience and perspective is an unavoidable issue. It concerned both of us during our collaboration. He frequently stressed that much of his experience was typical of his generation, and he always endeavoured to get the facts "right." But he understood that his experience and explanation of things might not be the last word. This is an established skipper's story. It is a Fortune Bay perspective, rooted in a small inshore fishing outport at the periphery of a region its people often described as Newfoundland's "forgotten coast." Other men would have somewhat different experiences and perspectives. Neither of us would encourage naive acceptance and over-generalization from every point made in this account.

It is reasonable to expect some differences in fishermen and industry experience due to region, individual circumstance, opportunities and choices made. For example, someone from a Placentia Bay outport, near Burin or Marystown, whether Irish Catholic or Protestant, perhaps from a shattered family, would tell another story. Yet the generational and occupational experience and basic understanding of the structure of the fishing industry, work, and community relationships would be quite similar. My discussions with many former south coast Newfoundland banks schooner and trawler fishermen, and their wives during the 1960s and 1970s, and library and archival information support the view that "he had done it all." However, a volume based upon many witnesses to Newfoundland's south coast fisheries experience would be an instructive future task.

This volume has several purposes: Arch and I shared the hope that his story would help younger generations better understand banking schooners and dory fishing, the groundfish trawler, their fishermen, and the transition from sail to power in our fisheries.

There are important issues here. Two examples will suffice. The first concerns the political voice of Newfoundland fishermen. When in 1967 I began study of the organization of work and life around Newfoundland's modern trawler fishery, its trawlermen were still unorganized "co-adventurers" in industry and government terms, as they were before in the banks schooner and dory era.[2] The deep sea sector had adopted new catching and sailing technology, yet its labour followed an exhausting, year-round, fully industrial work regime that dominated more of their lives than before. Familiar with other industrial occupations on the mainland, where organized labour was commonplace, I wondered how this extreme industrial domination was justified and accepted? Its trawlermen were reluctant to openly criticize or move to change their situation. It remained unchanged until the early 1970s. Arch's story, and my commentary suggest some of the answers.

A second issue involves what this account may contribute to the discourse about the present crisis in Newfoundland and eastern Canadian fishing industry. Arch's experience is part of the technological and corporate revolution in twentieth century Northwest Atlantic fisheries. It reflects global trends that compel attention. Unrestrained, over the years they have changed hundreds of thousands of lives once linked with the harvest of the sea's living resources. In Newfoundland, as in coastal hinterlands elsewhere, they have resulted in unprecedented fish stock destruction and the unemployment and displacement of tens of thousands of people. How did this happen? This account indicates that these destructive trends have long and complex historical roots. Responsibility for them runs deep and wide.

Further, Arch's account of struggles with life's challenges typical of his generation reveals values that helped its members meet and overcome them. As heritage and identity, this story offers a measure of confidence to younger generations of Newfoundlanders who today face equally, if not more challenging times and uncertainties. "If the old timers could make it with what little they had, surely we can make it."

Captain Thornhill's voice is the core of chapters one through eleven. Each of these eleven chapters is preceded by an introductory commentary. It contextualizes and highlights important issues in his first person account. Chapter twelve sums up the balance of Arch's career, offers views of the man as a representative skipper, and closes the saga. Endnotes offer selective historical links, supplementary information, and reading suggestions that some readers may find useful. A list of references, Appendices, and a glossary round out the volume.

What ya got, ya got

Finally, what follows is the "catch" from our first exploration of a new ground together. Had Arch lived to conclude our work we would have revisited this ground again and again for more information before our voyage was finished. But, this reminds me of what another old man observed years ago when he and blacksmith Wilson Osborne sat together, tea-mugs in hand, in the galley of a schooner. As she rocked at her mooring and wind driven halyards sounded a rythmic tattoo against her mast, over and over, he said, "Do ye hear what she's sayin'? Do ye hear it, boy?: What ya got, ya got. What ya got, ya got."

Like an old fellow said one time,
 "*When the dead rise again, there
 will be more rising from the sea in
 Fortune Bay than in any other place.*"

Chapter One

An Excellent Place for Fishing

This first chapter begins Skipper Arch's story. He recalls his youth early in this century in Anderson's Cove, a typical, small and remote fishing outport at the mouth of Long Harbour, Fortune Bay, on Newfoundland's south coast.[1] It was close on rich marine fisheries that first drew Europeans to these shores. The mouth of Long Harbour was the gateway to rich salmon runs. Before the 1870s Long Harbour was reportedly a "hunter's paradise" that attracted English and French warship officers, and sportsmen-aristocrats. It was also a highly productive area for herring bait used in the cod fishery.

Arch nostalgically recalls his early years when visiting American schooners arrived in search of herring. However, these visits were a vestige of more turbulent times. During the 1870s many American banks schooners visited the Cove and elsewhere in Fortune Bay for herring they either caught themselves or obtained from local fishermen. But soon the Anderson's Cove fishermen came to blows with the Americans, ostensibly over local fears that the herring stock would be overfished. Anger aroused, in 1878 local fishermen forced the Americans to cease their seining.

Anderson's Cove had perhaps ten or twelve families early in this century, and about twenty families and 100 people at its high point in the mid-1950s. It was totally abandoned in 1966. Life was intimate and quiet in the small

15

world of tight knit families, many related through the male line. Daily work and living routines were usually broken only by life cycle and spiritual-religious events —like birth, marriage, death, the twelve days of Christmas, and scheduled 'times.' These were **community** occasions of joy, sadness, and worship. They invoked, celebrated, and reconfirmed values and social relationships forged between the grindstones of daily living.

Arch's people had settled here to produce salt fish and other fish products destined for oversea markets they accessed through large merchant traders. Their lives pivoted around seasonal exploitation of coastal fish stocks — cod, herring, lobster, and, for both trade with merchants and household consumption needs, nearby land resources, like wood, berries, caribou, and other game. (Arch recalls the first time local men shipped out for bank fishing was only during the first decades of this century.) On a credit basis fish merchant traders advanced their fishermen-trader clients needed commodities they couldn't produce themselves, in exchange for fish. Little cash was exchanged and it was scarce locally.

The Anderson's Cove location was very rugged, it had little pastureland, yet most families had a few stock animals — cow, sheep, even pig. They did manage small vegetable gardens, mainly for potatoes and cabbage, and built their own homes. Everyone learned the importance of self-reliance, self-sufficiency, and sharing. What a household's members couldn't make they did without or had to obtain elsewhere, like flour, sugar, molasses and tobacco, and vegetables from Prince Edward Island, all obtained through fish merchants. Household survival hinged upon every member's efforts. Men and women, boys and girls, learned to make do, and acquired diverse skills essential to economic production. Children pitched in beside parents as soon as they were able. All feared and shunned the poverty and stigma of "the dole," or welfare dependency. But the mix of commercial and subsistence activities rewarded hard work with a meagre material existence.

1901-1910

Sometimes I lie in bed for hours in the night before I fall asleep. My mind goes back to the schooner days, the banks fishery, and I relive my life from the first schooner I was ever on right up to the last one. I see the schooners coming into Anderson's Cove with everything set — mains'l, fores'l, jib and jumbo, and gaff topsails. And I know all their names as if it happened yesterday. But it's all like a dream now.

The youngsters growing up today know nothing about that, the hard work, the hardships and risks, the tragedies and lives lost. Nobody had anything, and, yet, you never gave up struggling, trying. They are the ones who would be interested in what we did, and what we had to do. I want to make everything clear so they will understand.

Most of the men who went deep sea fishing from Newfoundland to the Grand Banks came from the small fishing settlements or outports around Fortune Bay, and some from other places around the south and east coasts. My father's grandfather, I suppose, came from England. And my father always said he was born where English Harbour West is today. His father used to fish off Anderson's Cove in the summertime, and he decided to make a home there, and that's where all my father's brothers were born.

No one lives there today

Anderson's Cove was one of the safest harbours in Fortune Bay. It probably got its name from someone who lived there or in a nearby hole years before my father settled there. No one lives there today. It is a small place at the mouth of Long Harbour, an excellent place for fishing. Just outside the Point it was excellent for herring. And its deep water is sufficient

Anderson's Cove, looking north up Long Harbour, c. 1960.

for almost any size ship. Many banking vessels went there for bait in the spring, late fall, and winter of the year. In earlier years man-of-war ships went within three or four miles of the harbour bottom, and the remaining distance in a steam launch, to reach one of the best salmon fishing rivers in Newfoundland.[2]

And it was an excellent place for wood, game and berries. There was always plenty of wild meat to get then, especially rabbits and partridge, and salt water birds. But caribou was the big thing. Our caribou hunting grounds were about eighteen or twenty miles from our home. Twelve miles of this distance had to be covered by water, up Long Harbour. But ten miles of this stretch would freeze up around the last week in December, and it was a real highway until late spring. We went over this stretch of ice with dogs and sleighs until we got to the harbour bottom, and then we went five or six miles into the country to get whatever caribou we wanted. I was born there, in Anderson's Cove, Fortune Bay, February 27, 1901, one of nine children — five brothers and four sisters. It was a quiet old place. There were only eight families there until I was twelve years old, and four of these were my dad's brothers. Then three families moved in from a nearby settlement, which meant another twelve or fourteen people. My boyhood was quiet. It's easy to see there wasn't much excitement.

Being able to write your name meant a lot to us

There were no teachers there then, and my father's family was uneducated except for Uncle Charlie, his youngest brother. They were unable to read or write.[3] Uncle Charlie walked from Anderson's Cove to Stone's Cove, about one and three-quarter miles away, in the mornings after breakfast, summer and winter, until he completed Number Three Reader at the Anglican school there. The people in these settlements were Anglicans.

Father and his brothers were very concerned about education. They didn't want their children — us — to grow up like they did. They were born and died of old age and couldn't write their own names. Had it not been for their dogged determination, it would have been a long time, if ever, before we could write our names. They approached the Anglican minister several times. He was stationed at Belleoram, about fifteen miles away. He came occasionally and held his service in my uncle's kitchen. Our families were getting larger, we were growing up just the same as our fathers, and my father and his three brothers thought it was about time a teacher was provided. They tried one more time to impress upon the minister the importance of sending a teacher for their children. But the answer was, "No. We can't do it. There are not enough people. We can't afford it."

At about the same time, Congregational missionaries from England established a church and school at Little Bay East, eleven miles away by water from my home. The brothers talked things over and decided that something had to be done soon, so one morning they got in their fishing boat, went to Little Bay East and discussed the situation with the Congregational minister there. They said, "If you will give us

a teacher, we will come over to the Congregational Church."
'Twas only right they should. The minister agreed, and the first teacher was stationed at Anderson's Cove in 1902.[4] So some of us could write our own names before we went to the Grand Banks. And, believe me, being able to write your name meant a lot to us.

A Congregational church was built at Anderson's Cove soon after the teacher came. Anderson's Cove had everything then, and it grew to have fourteen houses when families from Hoop Cove and other small places moved in. They couldn't get teachers to their communities. School was held in Uncle Charlie's house, in a little room, a bedroom, off the kitchen. The minister also slept there when he visited. The school remained until 1965 when all of the people moved away. Some say they were drove out by the government.

There was only one doctor for all of Fortune Bay

I well remember my first day at school. I was only four years old and it was in the spring. Some time during the winter a finger on my left hand became infected, and, after a while, blood poisoning set in. There was only one doctor for all of Fortune Bay at that time, and he was stationed at St. Jacques, over twenty miles by boat from my home.[5] He visited the settlements around the bay in his little boat when weather permitted. For some reason I could never figure out, my mother let me go to school one morning with my older brother. I didn't have my name on the register because I wasn't old enough. I looked out the window and saw the doctor's boat coming in the harbour. My dad came to the school shortly after, took me in his arms, and carried me home. He told me the doctor had come to cure my arm.

The doctor operated on my left arm, and my parents often told me that he wasn't very optimistic about saving it. You can still see the scars from the operation. I will never forget the medicine he prescribed. It was the brimstone used in the batteries at the telegraph office. My mother put it on my arm three times a day for a long time. It was brimstone alright![6] I can almost scream now when I think of it. But my arm was well and I was able to register when school opened in September.

Years later there were several cases of smallpox in Fortune Bay, and the same doctor came around to vaccinate people. When my turn came, he said, "Young man, you'll have to be vaccinated on your right arm."

It was like coming to a big city

Stone's Cove was the nearest community to us and much larger than Anderson's Cove. It was like coming to a big city from a small outport. There were thirty or forty families there, and two or three small stores with plenty of goods on their shelves. Six or seven deep sea fishing schooners operated from there. But it wasn't a safe harbour. It was right open to Fortune Bay, and when a storm came from the southwest, it could be dangerous. Many times men had to get out of their beds and put extra mooring gear on their vessels to keep them from going on the rocks. Sometimes they parted their moorings and were a total loss.

Once, just after a hurricane, I went to Stone's Cove for the first time with my father, and there I saw a forty-ton schooner up on the rocks, full of water, and all the men in the settlement around her. I wasn't very old at the time, but I can still remember the name. She was called the *Louie H*. She was refloated, repaired, and sailed for a number of years. But later, while tied up by the owner's wharf with a load of freight on board, a severe storm came from the northeast. The vessel parted her lines, drove out to sea, and was never heard of afterwards. Fortunately, no lives were lost.

It became like a big family

The Stone's Cove people were like strangers to us youngsters until we were about ten years old. It was quite a thing when I went up to somebody's house in Stone's Cove with my father. It was "Mr." this, and "Mrs." so and so. Sometimes we boys had an argument and a scattered fight when Anderson's

Cove boys would go down and meet Stone's Cove boys. But, as we got older, our people seemed very close, especially when you were in schooners together, fishing all season. It became like one family. Then some people from Stone's Cove married people from Anderson's Cove.

At Christmas time there were thirteen days when you wouldn't do much except have a dance, a concert every second night. We called it 'a time.' If there was a time at Anderson's Cove, they wouldn't have one at Stone's Cove. They'd come down here. If there was a time at Stone's Cove tomorrow night, why, we'd all be up there. After the concert you'd have a dance and people would be up until daylight.

I remember one winter night, after a dance, it was freezing and blowing the biggest kind of snow storm. When we left the hall one of my friends from Stone's Cove said, "Arch, you come over and sleep with me tonight." When we went to bed, I took off my shirt and laid it on a chair. We slept late that morning, and when I got up and tried to put my shirt on, I couldn't do it. It was so full of sweat from our dancing the night before that it had frozen solid. So he loaned me a shirt.

There was a path about one mile long between Anderson's Cove and Stone's Cove. It ran beside a pond. That's where we had sports in the winter. We had some ice boats there. My uncle and older cousins made one with sails, just like a real schooner. On Sundays you'd spend most of the day up around the pond. And sometimes, on moonlight nights, we'd get right at the top of the big hills and come down on a sleigh right across the pond.

There were sad times in our little places

There were sad times in our little places. When someone died it was a long time before people got over the loneliness. The last breath wouldn't be out of them before the blinds were drawn, dark blinds. They were kept down for about two weeks after a person was buried, and people wore black. That went on for years and years. In the fall of the year after the fishing season, some vessels always went to North Sydney,

Stone's Cove, c. 1930

Halifax, Prince Edward Island, and the U.S.A. for freight, such as coal, flour, produce, and lumber. When I was just seven years old the Stone's Cove schooner, *Columbine*, commanded by Captain James Tibbo, went to Prince Edward Island for a load of produce. When he returned Captain Tibbo first put in at Belleoram to discharge some cargo before sailing the last fifteen miles home to Stone's Cove. He had his little five-year-old son with him. The *Columbine* left for Stone's Cove just before dark on a Saturday evening. They were light in ballast, and nearly all their cargo was out by then.

There were no weather warnings in those days and a severe snow storm suddenly struck. On Sunday morning, when Captain Tibbo's family was getting ready for church, there were cabbages, turnips, potatoes, apples, a little boy's cap, and other debris right under the house, all squashed up to their doorstep. The vessel had run ashore and was lost with all hands, six lives, almost in the harbour entrance. It was one of the worst tragedies in the history of Anderson's Cove and Stone's Cove. The keel of the *Columbine* was later salvaged and the *Mamie and Mona* was built on it, and she fished for a number of years.[7]

In the fall of 1911 a bank fishing schooner named the *Marconi* sailed from Rencontre East, Fortune Bay, for ports in the U.S.A. She carried a crew of six including the captain. Two of the crew belonged to Anderson's Cove, and one belonged to a place with only two or three families about three-quarters of a mile from my home. They arrived safely at their destination and left again for their home port in the month of January. Time passed and no news was heard from the vessel. Eight days was the normal time for the vessel to arrive. People, especially the relatives, grew worried. January was slipping away and still no news. It was a very stormy month and the older people said the vessel didn't have a chance to stand up against the gales and frost.

Finally, the vessel was given up as lost with all her crew, the window shades were drawn, and a time of mourning began. I was only eleven years old at the time, but I can well remember the gloom cast over the little settlements. After a few days the men carried on their usual work, herring fishing and hunting in the country for caribou.

The boys are not lost!

Anderson's Cove was a very lonely place at this time, especially when you paid a visit to the parents of the two boys on board this long overdue vessel. One morning the father of these boys awoke and said to his wife, "The boys are not lost. I dreamt last night that they came home." The mother wasn't convinced. But, lo and behold, a few days later a telegram came saying the vessel had arrived at Trepassey and all the crew were safe and well. The vessel had come within reach of land several times with most of her sails gone. But it was driven off over the Grand Banks. All of their food supplies were gone. But the cargo fortunately consisted of foodstuffs, such as flour, sugar, and apples. They were short of water, so the men climbed to the mast heads to collect snow to melt.

It wasn't easy to get from Trepassey to Fortune Bay in winter, so the men had to wait a while for a passage home. Everybody prepared for their homecoming. It was to be one of the biggest events in the history of Anderson's Cove and Stone's Cove. As the three men, back from the grave, entered the harbour, there was such a burst of gun fire, flags flying and bells ringing! A day of rejoicing never to be forgotten, and, to add to the celebrations, there was a Grand Bank vessel named the *Bessie MacDonald* moored in the harbour to take on a load of herring. Her crew was firing the cannon and ringing the bell.

Suddenly, in the midst of the celebrations, Uncle Charlie, who was telegraph operator at the time, came down the path to the harbour with a downcast look. He brought sad news that a man from Stone's Cove had shot himself on board his vessel. This cast a gloom over both settlements on that day and for a long, long time.

This man was skipper of the *Mamie and Mona* and it was his first command. He went on board the vessel every day, making preparations for the winter fishery, which was to begin in a week or so. This evening, when it came time for him to be ashore, he didn't turn up. So another man working for the firm got into his dory, rowed over to where the *Mamie and Mona* was moored, and climbed aboard. He went into the foc'sle and called out his name, "Jim." No answer. Then he walked back aft to the cabin.

By this time it was getting dark outside. It was about pitch black in the cabin when he stepped inside. There wasn't a sound. He stepped across the cabin floor and tripped over the man. There was a gun alongside him, and a pile of blood. He had blown his brains out. This happened on my oldest brother's birthday. I could never forget it.

Only a few weeks after, another tragedy struck our two communities. This same man, although he had no learning and probably couldn't write his own name, had been chosen to skipper the *Mamie and Mona* when another man was given the new *Stanley and Frank* in place of Captain John Price.

John Price was a smart young fellow who left Brunette Island and settled in Stone's Cove. He went deep sea fishing on the banks in his early years and was a fine seaman and dory fisherman. A banking vessel owner at Stone's Cove decided he would make a good captain, so he offered him the fifty-ton *Mamie and Mona* to command, and he accepted. During the next two or three years he brought in many bumper trips from the Grand Banks. Then the owner decided to buy an eighty-ton vessel for him. She was built in Nova Scotia and named the *Stanley and Frank* after the owner's two sons.

Captain Price finished the fishing season in October, and brought the new vessel to Stone's Cove in the fall of 1911. She arrived with flags flying, guns firing and bells ringing, as was the custom when a new vessel was purchased. They moored the vessel and made plans to go winter fishing after the New Year.

After being home a few weeks, Captain Price and two men decided to go hunting up in Long Harbour. To be sure they would get away the next morning, they put their gear, sleigh and axes, food and guns in a dory with the intention of carrying it to Anderson's Cove. Men often did this because it was much easier to get out of Anderson's Cove than Stone's Cove on a stormy day. There was a strong wind when they left, although nothing like a storm. But as they rounded Stone's Point, always a bad place, only a quarter of a mile from their homes, suddenly a sea swept their dory and turned her over. Captain Price and Randall Pope were both drowned, while the other man got ashore on the rocks.

The captain had married a young Stone's Cove woman not long before and had one child. He was a member of the Stone's Cove Orange Society, so when his body was recovered they erected a monument in his memory beside their hall. He would have been among the most outstanding fishermen of his day had he lived.

You gave it more thought in the small places then

There were only about twenty-five or thirty people in Anderson's Cove then. When someone died like that in these small places, after the funeral it was so lonely you could hardly go from one place to another. Children, parents and even the elderly people, if they were going someplace after a funeral, By Jeepers, they always wanted company! The cemetery was across the harbour around the Point. When someone died, and you looked over at the cemetery at night, you'd say to yourself, "Yesterday he was here and there he is now, out on the Point." You gave it more thought in the small places then, more than you would today.

The few families of Anderson's Cove were all hard working people. Only one family there ever went on the dole, and that was because of sickness. The father was a sick man all his life and not far up in his thirties when he died. He left a large family, but when his boys grew up, I'm telling you, they didn't go on welfare.

Our people built and owned their own homes. They were different from the ones built today. They were tight, almost square, two-story houses. Most had a steep roof, but some had a flat roof. There was no central heating or running water. Houses were heated by a big wood and coal stove in the kitchen. A large family might have four bedrooms, but three was average. And there were no basements like we have today; just a little place big enough to crawl around in, dug out for a cellar under the house where we stored some vegetables like cabbage, and bottled foods like salmon and moose meat. But potatoes were usually stored in a big box or barrels in a back porch or in a locker somewhere in the house. Many times my mother wrapped quilts around the potato box in the winter to keep them from freezing. Most of these houses were not just houses; they were homes, and in spite of the lack of so many modern conveniences I remember my home as a happy one.

The older men were inshore fishermen

Out of the eight families at Anderson's Cove, only one man went bank fishing until the sons grew up. Then all the sons went. The older men were inshore fishermen. I remember when the younger men first "shipped out," as it was called.

The pleasant memory was not their leaving home, but their safe return in the fall of the year when the voyage was over. Most of them went in the Stone's Cove bankers at first. After a few years, they went to Belleoram and Grand Bank, and I was very anxious to become one of them.

There wasn't much farm land at either Anderson's Cove or Stone's Cove. Mostly rocks and cliffs all around. But there was always plenty of hard work to do to make a living. We fished as much as we could, the year round, except for blowy, stormy days. The lobster season opened in April month, and we fished until the caplin made their appearance and began rolling in on the beaches in the first or second week in June. The lobster pots wouldn't fish then because the lobsters were glutted on caplin, so we left them in the water and stopped lobstering until about the first of July, when the caplin run was over. Then we returned to lobster fishing until the 24th of July, the deadline to take our gear from the water.

Everyone grew enough vegetables to last until the schooners came back from Prince Edward Island in late November or early December with a load of produce. August month was the time for cutting grass and getting it dried and snugged away in hay lofts. Most people had a few cattle and sheep, which were very important to a family. We looked forward to the day to kill a lamb, a cow, or hog. The mothers would knit warm clothing from the sheep's wool.

We sold our caribou meat to the few wealthy families around

And everyone looked forward to going in the country in late August and early September to kill caribou and bring it out on your back. This was hard work, especially in the warm time of the year. I was thirteen when I first went in the country for caribou. I brought out fifty pounds on my back, which was good for a youngster. Some years later I got my own gun, but I never killed many caribou. I let the "sure shots" do it.

We sold our caribou meat to the few wealthy families around. After we arrived home, we first went to Stone's Cove to sell every pound we could. They were a little better off there because they had banking vessels and some business. It was a wealthy place compared to Anderson's Cove. They had enough to buy a few pounds of meat. A merchant might buy fifteen or twenty pounds and retail it out. But there were many people who would like to have it but couldn't buy it. They never had a cent in their houses.

Then we got in our dories and went ten or twelve miles around Fortune Bay to sell from door to door. There was very little left for one's self. The meat was delicious, but this had to be done to get money to buy our three or four tons of coal, which was brought by schooner from North Sydney, and vegetables from Prince Edward Island to see us through the winter. And there was that five-dollar bill my father was good and sure would be tucked away for the church. He believed that if you didn't have that five dollars for the minister, you wouldn't get any fish next year.

We also hunted caribou later in the year, just before the Long Harbour freeze-up. Two Anderson's Cove families had winter houses at the bottom of Long Harbour. When one, the Riggs family, moved away from Anderson's Cove, the people from home and Stone's Cove who went up in Long Harbour to hunt in the winter bought Riggs' old house to stay in at night. It was always called the "Old House."

People from other places, like Femme, Harbour Mille, and Little Bay East, around the bottom of Fortune Bay, all stayed there. Sometimes there were fifty or sixty in the house at night. Next morning, they'd take off into the country and be in there four or five days. You would go up twelve miles in dories, and then several miles into the country. The weather was cold enough that it would be freezing in the country, so, after you shot what caribou you wanted, you bulked it up. Two or three weeks after that, Long Harbour was frozen over hard enough you could run around on it. Then you'd go back up with the dogs and sleighs and tow out the caribou killed weeks before.

Everything from a needle to an anchor was obtained on credit

Our people and Stone's Cove people sold much of this caribou to a man in Stone's Cove, and he sold a lot to merchants in Belleoram. Nearly all the northsiders, the schooners on the northern side of Fortune Bay from places like Stone's Cove, Belleoram, Harbour Breton, English Harbour and St. Jacques, went winter fishing on Rose Blanche Bank in January and February.

The merchants bought a lot of caribou for their crews of twenty-odd men. But at this time they wouldn't give you cash for it. You had to take it up in goods. That's how my father got his first rifle. He killed a caribou and carried it up to Stone's Cove, sold it, and got a rifle. Cash was very scarce these times, and everything from a needle to an anchor was obtained on credit. And, believe me, the merchants charged plenty for their goods in those days.

Handlining for cod was another important activity in August. We used a single, small line to the bottom with two baited fish hooks. About the first week of September we began fall fishing using manilla twine cod nets — what we now call gill nets — and trawls with squid for bait.

August month was always a busy month for everyone, old and young. There was no such thing as selling fish fresh these times. All the cod was split, cleaned and salted. After laying in salt for a time it was washed and put on a bedding of fir and spruce boughs spread over flakes made of long sticks called 'longers.' The mother and children in each family did most of the fishmaking for their men by laying it in the sun. At Stone's Cove, women cured fish from the schooners for extra cash. The same was done at other places, like Belleoram, St. Jacques, and English Harbour West, when schooners landed fish.

There wasn't anything wasted from the cod fish these days

The women always had charge of the cod liver when it was taken from the fish while splitting, which means removing the 'sound' or backbone from it. They put it in barrels and punchions where the sun brought the oil from the liver. You would see them go for the blubber barrels with a small container almost every sun blast, to skim every ounce of oil they could possibly get. I can assure you that there wasn't anything wasted from the cod fish these days. Fish roes, or cod roes, would be salted and sold. The small children loved them. Then there was the 'sound' taken from the 'sound bone' and salted for food, and the heads of the fish were salted in barrels for food during the winter months.

At different times each year merchants came around in their little traders to collect the fish when it was fully dried. Uncle Charlie, in Anderson's Cove, bought some, as did other merchants from Belleoram, English Harbour West, and Bay L'Argent. One man might get his credit from one merchant, and the next man from someone else. The price they said they could afford to pay was only $2 or $4 a quintal (one quintal is 112 pounds of cured fish), there was nothing you could do about it — only accept it.

After the fish was weighed and stowed away in the boat you went in the little cabin to settle up. Then, if there was a dollar left after all the spring and summer expenses for your gear, rubber boots and clothes, and salt, etc., were paid, you took up the worth of it in food and clothing.

The things women bought from their work kept their homes going

The women got from fifteen to twenty cents per gallon for cod liver oil, and bought many useful articles with their money's worth, such as paper and paint for the inside of the house, and curtains and clothes for the children. And bed linen on end. The things women bought from their work kept their homes going and didn't have to come out of father's earnings. Some had big families and they couldn't make a go of it just by fishing.

Many long hours and days went by waiting for the little traders to approach the harbour. They arrived loaded with provisions, salt, fishing gear, and other goods, and they took a load of fish and oil, etc., back with them. They didn't leave much cash behind, but the goods were surely appreciated. I well remember Christmas time, when we were lucky to get a five- or ten-cent piece to change for coppers and pennies to make it jingle in our pockets. But money wasn't always the most important thing. There were happy homes.

I often think about the berry-picking season. The little traders always came around a couple of weeks before the season started, with a deck load of wooden barrels. Each barrel held twenty-eight gallons. They were distributed to whoever wanted them. Partridgeberries and blueberries were picked mainly by mothers and the children who were old enough.

A reliable woman selected her own crew of five or six, two or three weeks before the berries were ready to pick. Three or four such women and crews each set out about the same time in an old dory, usually one condemned from fishing. To be there by daylight, they had to rise from their beds around two o'clock in the morning, do their housework and feed the younger children their breakfast. The youngest children were left in the charge of an older one, which wasn't very old. They rowed about six or seven miles up Long Harbour. They often couldn't keep time with the oars and, owing to a leaky bottom, they'd be bailing water from the time they left home until they reached their destination and home again. But they'd get there just the same.

Believe me, it wasn't always sunshine, and many a tragedy was only narrowly avoided. Seven miles was a long distance for women to travel back and forth in a dory, and they often had to shelter all night for the strong wind. Most of the time, when the mothers and children were berry picking, the fathers and older boys were fishing, with the exception of a windy day, and then they would join the women and go berry picking. The women weren't sorry about that on windy days. Then they had no responsibility for getting back and forth.

When the season was finished, the traders returned to collect the berries. They paid from ten to fifteen cents per gallon. An average day's work would yield six or seven gallons per person. The women and children had to do the same with their berries as the men did with their fish: take up the worth of it. Not a cent in cash would one get.

There was scarcely any cod fishing done in October and November month as the weather was mostly stormy all the time. For a few weeks, men and young boys from Anderson's Cove and other places around went up Long Harbour in their dories to cut wood for their winter and summer needs. I never liked that. We had to carry it on our shoulders from the woods down to the shore. Sometimes I had an inch of dead skin on my shoulders afterwards. But it had to be done.

Herring was in demand all over Newfoundland

The herring fishing began around the last week in November or the first of December. That was always a happy occasion. Herring was in demand all over Newfoundland. Banking vessels came for bait in December, January, and in the spring of the year. Small boats came from all parts of Fortune Bay to catch and sell herring. Sometimes the harbour was filled from one end to the other with vessels, mostly from Gloucester, and "Dutchmen" from Lunenburg. In the late 1890s French vessels also came for bait. Sometimes you could walk across the harbour on them. Yes, Anderson's Cove was a busy place in the winter months in the early 1900s, despite its small size. We had few fishermen, but sometimes the harbour was filled to capacity.

The herring was caught, then frozen with natural frost, and there was plenty of it in those days. American vessels brought lumber, and their men built scaffolds almost to the top of their masts. The herring our fishermen couldn't freeze on their own wharves was sold to the vessels and frozen on the scaffolds.

Herring was caught with gill nets with about two-and-a-half to three-inch mesh. And there were seine boats, approximately thirty or thirty-five feet long, which used a purse seine to take the herring from deep water. If the herring came in November, it was often salted in barrels and sometimes in bulk. Some winters the herring would be plentiful, and sometimes there wouldn't be any at all. Many times the vessels came from Gloucester and returned with no cargo.

Once loaded, the vessels were very deep in the water, with hardly any freeboard at all. It was a dangerous situation. Many were lost with all hands on their return trips to Gloucester. I wasn't very old when one vessel left Woody Island in the Bay of Islands, on the northwest coast of Newfoundland, with a full load of salt bulk herring bound for Gloucester. A severe storm came on the first night out and the vessel was lost with all hands. The vessel and crew belonged to Woody Island, and most were close relatives with the same last name: Hackett. The vessel was commanded by a Captain Hackett as well. The last of these tragedies occurred in March, 1935, when the schooner *Arthur D. Storey* and all her crew were lost going from Fortune Bay to Gloucester with a load of herring.

Everyone looked forward to the Americans' arrival because they paid mostly in gold for their herring.[8] It was about the only time people saw cash in any quantity from one winter to another. But you never knew when they'd arrive until they came around the Point with their beautiful, white sails shining. What a sight! It meant a lot to the people in many ways. They brought many nice things, and we children, especially, were very anxious to see them. As long as they were in our harbour we were always sure of apples and good food when we went on board, mostly right after school hours. And the lumber used to build scaffolds on their masts was always given away when their vessels were loaded and ready to leave. It came in handy when people were building houses both in Anderson's Cove and Stone's Cove.

Chapter Two

I Never Gave Up Once in My Life

In this chapter Arch looks back to the second decade of this century in Anderson's Cove. Fortune Bay outports and offshore vessel centres appear beehives of economic activity. A growing and diversified fishing industry inspired bouyant optimism. Ambitious fishermen, like Captain Thomas Bond of nearby Stone's Cove, were able to marshal the capital and skills to build or buy vessels, even to become merchant-schooner operators. Most remained agents of supplier-creditors, but their success demonstrated that hard work and initiative pay.

Like hundreds of other outports, Anderson's Cove was on a frontier with still abundant, yet increasingly pressured fish resources harvested for far flung commercial interests then centred at St. John's, and in West England. From south coast bases like Harbour Breton, Belleoram, and Grand Bank, fish merchants advanced provisions to fishermen on credit — on their terms — in exchange for fish to supply their markets in Europe, the Caribbean, and South America. This chapter only hints at how this trade relationship enabled merchant managers and vessel owners to harness and dominate outport and fisherman lives. It is more explicit later.

Knowledge, skills, and values

Newfoundland's outport fishing households followed year-round routines keyed to extraction of a mix of fish, game, and forest resources for commercial and/or family consumption uses. Most children were initiated into the ways of their community and environment with a seamless ease. Their play dissolved into productive household and community roles with a continuity and clarity that many term "traditional," as parents, relatives, and neighbours, directly and informally infused **their** children with essential life skills and knowledge about their social world, work and community roles and activities, the land and sea, and moral-spiritual matters. Blended together, they fostered strong personal initiative, and identification with family and community, environment, and Christian faith.

A sense of inequity and a better life was an important part of their basic social knowledge. Outports were not simply isolated self-reliant communities of working class households. Their contacts with Newfoundland's privileged classes, and with outside societies — especially Nova Scotia and New England, through merchant seafarers, fishermen, and sportsmen, sharpened awareness of disadvantage and exploitation in their economic relationships and belief that a better life was possible — if not here, then elsewhere.

Outport work roles were largely divided along gender lines. Children followed their parents' example into dawn to dusk labour that contributed to mutual well-being. They began knitting twine in the kitchen at an early age. Girls concentrated on household and family management skills, gardens, making and repairing clothing, care of stock animals where present, drying fish on shore, and family care. Boys were expected to gravitate toward fishing and other outdoor skills with commercial and subsistence production outcomes, and they built and maintained the home and their premises. Arch began fishing at age ten with his father and older brother. Formal education was only important to a point. Annually, at the highpoint of fishing, children commonly absented school to fish/and or work beside their parents. The productive roles and activities open to youth encouraged a strong sense of belonging to family and their community. They fostered strong social relationships while other customs restrained the potentially divisive, individualizing and class forces unleashed in larger communities. They encouraged an eagerness to learn skills that made one versatile and adaptable, an uncomplaining willingness to pitch in and work hard, and to share results with their families.

Outport life also built broad knowledge of local industry, its skills and values, and, as with young Arch, self confidence and a reputation as willing and able bodied. These ways and values also prepared young men and women for marine- and other occupations, whether in Newfoundland, Canada, or the United States.

Tradition at work

The success of seasonal economic activities also hinged upon understanding and anticipating nature's resource rhythms and uncertainties. This practical and often implicit insight or "traditional environmental knowledge," is familiar in current debate about fisheries and their problems, especially as it relates to marine ecology and the sea's living resources. However, it is poorly understood, and Arch's story gives only fleeting details about the marine ecology as south coast fishermen understood it early in this century.

By contrast, he gives frequent comment on events of hardship and tragedy involving seafarers and fishermen. Thus a father and son and crew were lost rounding Long Harbour Point; and mother's kin were lost off southwest Newfoundland. In short, generations of one's kin and others hunted these same lands, harvested its forests, and sailed and fished its waters. These tales hold more than meets the eye.

Like local ballads, poems and ghost stories about land and sea, they have many facets: They acknowledge the grief that often follows seafaring; they offer practical models for behaviour and lessons; and they invest turf and fishing grounds with personal spiritual associations that ease one's connection to environments that may seem otherwise indifferent, impersonal, and threatening. In result, one may feel less alone, even an intense sense that one's life is inseparable from the sea and nature. This is tradition, a heritage at work. To be sure, however, the sea's indifferent cruelties drive many away from its occupations.

An extra "leg" to stand on

Outport life early in this century gave abundant demonstrations of life's hardships, uncertainties, fragility, and inequities. For example, illness and accidents often made the transition to adulthood abrubt and stressful. A father lost at sea compelled many a boy of twelve or thirteen years age to assume an adult role to support his family. Likewise, the eldest daughter of a mother lost to illness or childbirth often left school to manage her father's household and its younger children or to work away, perhaps "in service" as a domestic with another family. And meagre returns from fishing and other household efforts to be self-reliant and self-sufficient drove many families to starvation's brink with little recourse for help from outside one's own community.

Although expressed more implicitly than explicitly in Arch's story, such unpredictable calamities and circumstances stirred a strong spiritual consciousness in his people. As with Arch's father, every household laid aside $5 for the church each year . . . to insure next year's catch. Denominational faith, prayers, and keeping the Sabbath provided an extra "leg" to stand on. Hence the often met notion that religion was at the "centre" of outport life.

Inshore vs. offshore fishing

Transparent contrasts between inshore and deep sea fishing were among the factors that stirred aspirations for opportunity beyond the small outport. The two were already polarised and sometimes in outright conflict on the fishing grounds. The inshore, immediate and familiar, spread its uncertain outcomes over several species and meshed well with onshore seasonal subsistence activities (e.g. hunting, trapping, and logging). Risks to life and limb seemed reasonable, less threatening. It meant home, family, and community. It was continuity. Yet it was marginal and precarious. The onset of World War I and loss of key European lobster markets made it more so.

By contrast, although it depended mainly upon one species, cod, the mobility of banks schooner fishing promised, without certainty, cash incomes beyond what Arch's father could hope for from inshore fishing, and the excitement of distant fishing grounds and ports. It meant a fundamental break from the inshore lifestyle and absence from home and immediate family. It was a large step toward the year-round occupation that came with groundfish trawlers in the 1940s.

The different work, hazard, lifestyle and family life implications of these two paths are evident in Arch's recollection of pressure to remain home, helping his father and uncles inshore, against the pull of banks schooner fishing. In the end, his parents' resistance only delayed Arch's departure. He knew he was lucky. Others were catapulted into it reluctantly. And he felt himself ready.

Many things had prepared him for banks fishing: A youthful initiation into the ways and environment of his community was infused with images and knowledge of the sea. He had learned traditions, lore, and skills of seafaring from older men, family and local role models, who faced its challenges. While his parents feared the hardships and dangers, perhaps inexperience and self-confidence helped inure him to them. From an early age its vessels, catch and income prospects, and the idea of becoming a skipper himself drew him offshore. He fantasized about becoming "someone," like his older cousins who skippered schooners.

A 'chance' doesn't come easy

But not everyone was given a 'chance' to test their ambition and skills on the banks. How to get it? First, Arch reached for the lowest rung on a schooner's ladder, to be deckhand or "kedgie." Although given a chance through a skipper uncle, he was forced to abandon it to ease his mother's fears. He didn't "have" to join a banker to help his family, at least not then. Yet the desire burned on and a year later he went for the second rung on the ladder, to be dorymate with an experienced dory skipper and neighbour. With relentless persistence, he bargained for and bought his chance for $60. Did he know what he would face?

1911-1918

When I reached ten years of age I began fishing with my father. My older brother was two years older than I, and father would take the two of us with him. My older cousins were young men by that time. I remember when they'd arrive home from the banks. I'd say to my mother and father, "I'll be some glad when I get old enough to go to the banks." And mother always said, "My son, you will die." I could hardly cross the harbour in a dory without getting seasick. But I never gave up once in my life. That didn't keep me from wishing, especially when I saw the schooners coming in the harbour.

There was always a chance they'd hire a boy

When the Stone's Cove fishing vessels arrived from the banks with their catch of fish, it was taken from them in dories and rowed to the shore, then washed and carried to the flakes. There was almost always an early start in the morning, and the women and older children were on the flakes at sunrise to get it dried. The rule usually was, say, if a vessel carried four dories (eight men) or ten dories (twenty men), however many it carried, that's how many dory-loads of fish were washed in one day. The day's work was finished when this amount was washed and carried to the flakes.

When a boy was big enough to hold a fish up in his hand, he'd be looking for work. By the time we boys were eleven or twelve we were always there when a vessel arrived. I'd go aboard so quick as a schooner came in Stone's Cove, to see if they wanted any lumpers. And, natural enough, many of the men preferred to have two or three days off for a little rest, or to do some work at home, so there was always a chance they'd hire a boy to wash out the fish, to go lumper. They probably preferred men, but men were scarce because they'd be on schooners or fishing on their gear inshore.

I was always eager to help my parents

They paid us lumpers fifty cents a day, a lot of money in those days, whether we were there from daylight 'til dark or only four or five hours, and you got fed as well. My father always gave me his consent to go and look for the work, even though it meant harder work for him to row around to his fishing gear.

We'd probably be washing out that schooner's fish for five or six days until it was all out. One of her men might take two or three days off, and then another would be waiting to do the same when he returned. So we'd stay on that schooner until the fish were finished.

All the crew worked together washing out that fish. If you finished by eleven, twelve or two o'clock that day, the work was finished. But sometimes the fish was very hard to wash. It might have been in the vessel longer, and the salt was dried in. But we usually finished by 2 p.m. and then we had the day for ourselves. And fifty cents in those times — 50¢ pieces! Boy, oh boy! If you worked six days, you had six — $3.00 in 50¢ pieces! We carried it home and gave it to our parents. I was always very eager to help my parents in any way I could, and I always carried my little sum of money to mother.

There were five or six banking vessels fishing out of Stone's Cove, and not enough room for all the fish to be sun dried there, especially when they had big 'trips.' So some vessels were sent to Little Bay East to wash out their trips and get it dried. I was always on the lookout for such transfers. It meant a trip on the vessels, and probably four or five days of work. But there were times when I was glad to reach our destination after only twelve miles of sailing, I used to be so seasick. Sometimes, just as all the fish were washed in the vessel I was on, another one arrived at Little Bay East, and that meant another four or five days' work. I was almost always sure of the jobs.

I have many pleasant memories of Little Bay East. My mother came from there, so we visited it many times in our young days. It was from there that I made my very first trip on a small sailing boat, my uncle's little trader, and my first trip on a banking vessel at the age of eight. The vessel was the *Sanuand*, a twenty-ton, two-dory fishing vessel from Stone's Cove.[1] I well remember how one of the crew members gave me the vessel's wheel to steer her along. My mother was going to visit her family at the time, made the trip in this vessel, and took some of us along with her. And I was at Little Bay East when I made my first decision at the age of twenty-six to go captain.

Little Bay East was a safe harbour and a nice looking spot. When I was twelve years old the population was between two and three hundred. Everybody had nice houses, quite a bit of land, and they kept cattle, sheep and goats. In those years quite a few men owned and commanded their vessels. Most were small and carried crews of from eight to ten men. Many came to this place to wash the fish they caught on the Grand Banks and off the southern Labrador Coast. The women and children "made" the fish on the fish flakes all along the water front. They used the dollars they earned from this work to beautify their homes.

Tragedy often happened

Tragedy often happened here as at other places. My mother's brother was captain of his own little schooner. He had his brother with him while fishing on the Labrador Coast, and the brother and his dorymate were drowned on their way to the mothership. My grandfather was visiting our home at Anderson's Cove when we received the sad news. My mother also had cousins who owned their own vessel, and one was skipper. They were bound for Prince Edward Island for a load of produce after the fishing season was over. They took shelter from a storm somewhere on the coast of western Newfoundland. While out in the dory, putting an extra anchor out, the dory capsized and two of the skipper's brothers lost their lives.

When I was eleven years of age I left school in April to go fishing with my father, and I worked with him until October when the weather began to get colder and I went back to school. In those days as soon as a boy reached the age of eleven or twelve, and was big enough to hold the oars in the dory, he would go fishing with his father and return to school in the late fall and winter months.

I began to feel I was somebody

My Uncle Charlie had a full-time job for three or four months as game warden over the Long Harbour River. When I was twelve he asked my father if I could go with him on the river. He gave his permission. This was the year the engines appeared on the scene. Uncle Charlie had a new, three-horsepower Imperial engine installed in his dory, and later, he became agent for these engines.[2] I began my new job about the first of May, and what an enjoyable summer we had. Long Harbour was ten or twelve miles long from our home, and we had many a fine run going there with the sail on the dory and the engine going. He taught me how to run the engine, and I was engineer most of the time. I began to feel I was somebody.

To make the summer more enjoyable, there was another boy from Stone's Cove working with his grandfather, commercial salmon fishing with nets. We all stayed at the same camp and we did quite a bit of trout and salmon fishing with the hook. We had to go quite a distance in the country to patrol the river. And we usually had plenty of company, because many sportsmen from St. John's, Canada, and the U.S.A. came to fish in Long Harbour River in the early years of this century.[3] When work was finished on the river I continued working with my uncle until October.

Uncle Charlie had many setbacks

Uncle Charlie also carried the mail to a couple of settlements named Conne and Femme. The mail boat arrived at our home once a week, and, weather permitting, we left immediately for these settlements. A distance of five or six miles, the little engine was a big help. We had it stormy at times, especially when the wind blew in through Fortune Bay, and many times we were glad to get back to smooth water.

Uncle Charlie had many setbacks and plenty of trouble in his lifetime. He was a good carpenter, and he started a small business and became the "king" of Anderson's Cove for a time. First, he built a small boat in which to go trading around the bay, carrying canned lobsters, salt, and goods, and brought back dry salted fish to sell to the bigger merchants.[4] He called the boat *Elsie* after his daughter.[5] The business lasted perhaps ten years. People often didn't pay their debts, so he had to close down.[6] But I heard him say with pride that, over the years, he paid every cent he owed to his creditors in St. John's.

Uncle Charlie's family endured much sadness. Their first child, a little girl, died. After that, three boys and another daughter made up the family, but joy soon turned to sorrow when the three sons passed away in their early teens, with only a few short years between. The eldest never reached his sixteenth birthday. The youngest son was one year older than I. My uncle carried on his business only for a year or so after his last son died. The daughter lived through this sad period. She married and had two little girls. Then, at the age of 34, she had an appendicitis attack and was rushed to Belleoram, where the only doctor in Fortune Bay resided. She passed away shortly after the operation.

He went . . . 'kedjie' because he wasn't old enough to go in dory

When I was thirteen years of age and my brother (Eli James) was fifteen, he left home for the first time in the month of January to go deep sea fishing with my mother's brother, Captain Wilson Green, of the five-dory schooner *Mamie and Mona*. He went as a deckhand or 'kedgie' because he wasn't old enough to go in a dory.[7] But during the winter fishing on the Western Shore, off Rose Blanche, one of the men became ill, so he had to take his place. He was on board only a few weeks when he became a full-fledged crew member, and carried on for the rest of the voyage. I well remember when he left home for the first time. It was like a funeral. I think I felt it more than the rest because we were always very close. It was no easy task for a boy of fifteen to go winter fishing, which was quite different from home-shore fishing.

Many of my cousins were deep sea fishing by this time and some a few years earlier. Two of them, twin brothers, became captains at an early age. The vessels began fishing early in January and returned from their first trip around the middle of March. Winter fishing was the most hazardous and hardest of the whole year, and it was always a relief to all on board when the trip was finished. My brother was in a vessel from Stone's Cove, and it called at her home port on the way to Little Bay East, where the fish was washed and put on the flakes to dry.

What must I do to become captain?

I accompanied him on this vessel to Little Bay East, and became lumper for one of the men who wanted a few days off. We had a smooth trip down the bay and I didn't get seasick. I wasn't on the deck very long before I asked if I could take the wheel and help steer the ship along. The man at the wheel said I could, and soon the skipper came along and asked how I was doing. I looked up at him and said, "Skipper, what must I do to become a captain?" It was in my mind to be skipper ever since I was twelve. I would see the schooners coming in with their sails shining, and go aboard. And I would see the skipper and notice how he stood above the rest of the crew. When my older cousin came home off a trip, I asked him all kinds of questions about skippers, how you became one and why they wanted to be skippers.

It was very cold washing fish, but the work had to go on. Just as this vessel's load was finished another vessel came, and I was assured of a job for another few days. There were five vessels fishing out of Stone's Cove that year (1913) and they arrived from their trips within a few days of each other.

In April all the fishing vessels were away on their second or spring trip. It began to get very busy all along the Newfoundland shoreline. From Cape Race to Cape Ray, people prepared for the shore fishery, mostly lobster and herring fishing. Most of the herring fishery was carried on in Fortune Bay.

There was always plenty of work for everybody

My brother had gone deep sea fishing, so I had to take hold where he left off, and that meant a full-time job. There was plenty of work to be done repairing lobster pots, going into the woods to get material to make new ones, and knitting twine to make cod and salmon nets. We began knitting twine at an early age. So many fathoms of twine were put out for us to knit, and it was up to us when we did it, before or after school, but it had to be done. There was always plenty of work for everybody.

The lobster pots were put in the water one day before the season opened, but you weren't allowed to bait them. This meant you had to have your bait secured a couple of days before the official opening. When the day came to bait the pots there was plenty of activity. The fishing gear was left in the water overnight before being hauled. It was always chilly and the water was very cold in April because of the ice melting from the rivers and bays. But we didn't seem to mind it too much. It was fish we were after, and that was our livelihood.

One lobster pot was always much larger than the others. It was moored a little distance from the water's edge, and when we arrived with our catch the lobsters were placed in this car-pot and kept alive for a week. Then came "crack-out day." That day, before leaving for the fishing ground, we pulled the large car-pot to the high water mark, and then the mothers took charge of cooking the lobsters. We boys were usually left behind that morning to help them, and we were always happy because it meant another couple of hours to sleep. Our fathers often left for the fishing grounds when the stars were still shining in the sky.

There was a little building called the factory, with a fireplace built within, and a large galvanized boiler placed on it. My work, this particular morning, was to get everything ready while mother finished her household chores. These days bring pleasant memories and I bet there wasn't a youngster in the Cove who didn't know and look forward to 'crack-out' day. Sometimes, when lobsters were plentiful, as they were in the year 1914, we had to crack-out twice a week.

The boiler held from thirty to forty pounds, and the lobsters were cooked about twenty minutes. After the first cooking, mother and the older children took the meat from the shells. The mothers always placed the meat into the one-pound and half-pound cans. It was afternoon before this work was finished, and when the men arrived from the fishing grounds, they secured the tins with solder. By the time the lobsters were bathed, cooked, and canned, and the factory thoroughly cleaned, there wasn't much of the day left, and our parents were often very tired.

The last week in May the Stone's Cove vessels returned,

and everyone looked forward to seeing their relatives again, and I looked forward to another job as lumper.

The war clouds began to hover around, and from the news flashes it was beginning to look serious. Many of us in these little places knew very little about war. Now and then we might hear of a fellow joining the British Navy. My mother's young brother was one of them. After a few weeks every news bulletin became more serious and the situation looked grave.

The early deep sea fishing vessels . . . made a beautiful picture

The early deep sea fishing vessels out of Fortune Bay had from four to six dories, and carried a crew of ten to fifteen men. By the time I was thirteen years of age larger vessels began to be purchased. Most vessels were painted black or green, and, with their sparkling white sails, they made a beautiful picture. Some vessels had two topmasts, and others one attached to the masthead. The topmast ranged from thirty to forty-five feet long, depending on the size of the ship. They all carried the "four lowers"; the mainsail, foresail, jumbo and jib. Those that carried one topmast had the gaff topsail, topmast staysail and a foregaff topsail, and a jib balloon. The sails on the topmasts were always called the "light sail." If the wind was increasing, the captain would say, "Take in your light sail, boys."

Most vessels arrived from their spring trips around the first of June each year. Some had bumper trips, others not quite so much. But, for most, it was a satisfactory trip. We boys were always glad to hear of the big trips, because it meant more lumper days for us. After the fish was discharged, and salt and food supplies taken on board, the caplin would be around. Each vessel had its seine on board and the crew secured their own bait, from ten to twelve dory-loads, depending on the size of the vessel. They secured the bait in four bins in the hold of the vessel with sufficient ice on it to last eight or ten days. The ice was cut from ponds in the winter and packed in ice houses and secured with sawdust until the summer.

Around the eighth or tenth of June all the vessels were on the Grand Banks fishing their caplin trips. The caplin season was always a busy one, as it lasted only about three weeks. This meant many long hours of fishing, both on the banks and at home, shore fishing.

The lobster fishery was discontinued while the caplin were around. On the 24th of July all lobster gear had to be taken out of the water. The lobster catch this year, 1914, was one of the best around our way, but everybody was worried because it looked more like war every day, and all the lobsters were exported to Germany. The vessels began to arrive from the caplin trips, and by the first of August they were all home, and some had left for their fall trips.

When the war broke out . . . practically all credit stopped for the lobstermen

I was at Little Bay East that August when the war broke out, and what a sad day it was for us and people all over the world. Our parents were really worried. Here we were with our whole summer's work of twenty-five to thirty cases of lobsters in the factory, and unable to ship and sell them. The only thing we had to depend on then to buy food was the small amount of codfish we had to sell, the lobster fishing being our biggest industry. We thought we were going to starve then, for sure, with nothing to eat but salt fish, no potatoes, no butter.

It was a serious situation for my uncle, because buying and selling lobsters was the biggest part of his business. Most of his suppliers could not pay their debts. The war broke out in August, the same month we usually shipped our lobsters, so they could not sell theirs. As the days went by the situation all over Fortune Bay looked very grim for the lobster fishermen, and for many of the other shore fishermen who depended heavily on their lobster catch. Practically all credit stopped for the lobster fishermen.

By the first week in October almost all the deep sea fishermen were finished with the bank fishery and most of the vessels went freighting to P.E.I. for produce, North Sydney for coal, and Halifax and parts of the U.S.A. for other commodities. My brother arrived home from his first deep sea fishing voyage, which was a successful one. His earnings were a great help to our parents and family in getting supplies for the winter. He was glad to be able to put every cent of his summer's wages, after his expenses were paid, into father's hand. They were very happy to have such help in these rough times.

Then the men from home and the different settlements around got busy getting their supply of wood for fuel, to see them through the winter months, and went hunting to get all the fresh wild meat they could. Others got their herring fishing gear ready. They had little more than one month to complete this work before the herring put in their appearance. Everybody looked forward to this, and by the last of November the herring fishery was in full swing, and a number of American vessels had arrived. It was very frosty after the middle of December, and the American vessels had their loads of frozen herring secured by the first week in January, and left for Gloucester, U.S.A. Long Harbour was frozen over this particular year by the last of December, and the Americans did some hunting for caribou before they left for home.

If you didn't get the fish you didn't make the money

My father was never a deep sea fisherman. He was never away from home for long periods in his life. But, owing to the serious problems in the lobster fishery brought on by the war, he saw no way to support his large family except by going to the bank fishery. He was over fifty, and it wasn't easy for him to do. And one was never sure of making a successful voyage because, if you didn't get the fish, you didn't make the money. He got in touch with a neighbour and asked him to take him in his dory; this he did. In the meantime, my older brother went winter fishing again on the *Mamie and Mona* with the same captain.

The kedgie . . . often made almost as much as the men

In February, 1915, I was fourteen years of age and had been fishing at home ever since I was eleven years old, so I saw no reason not to try to get a chance as deckhand or "kedgie," as it was called, on one of the Stone's Cove vessels. I wouldn't make as much money as a man in dory, but it would be a great help to my family. The kedgie got a guarantee of $25 or $30 for the season and the value of whatever fish he caught from the side of the vessel. The tails were cut off before the fish was salted, so it could be identified as the kedgie's when the trip was discharged. If on any day a member of the crew got sick and couldn't go in dory, the kedgie took his place and got a share of that day's catch of fish. It often happened, if the boy was a hard worker, that he made almost as much as the men.

The kedgie also salvaged the heads of the fish and cut out the tongues and sounds from the bone of the fish before it was thrown overboard with the rest of the offal. This was washed, cleaned and salted in tubs and barrels. Tongues and sounds were in great demand in those days. The kedgie could also count on a few extra dollars as lumper when the vessel arrived home.

I painted this picture as well as I could and gave it to my mother and father during the winter. Needless to say, I knew what the results would be, but I didn't give up and there was still a little while before the vessels returned from their winter trips. I kept insisting what a help I would be to the family, but my father said, "You can still help the family if you stay home and go fishing with one of your uncles." And mother said, "Your older brother has only been away from home one year, and your father is going for the first time, so how can I ever give my consent for you to leave home, especially when you are so young?"

But I kept talking about it and giving all the reasons I

could think of, why I should go, with the result that, about a week before the vessels arrived, they reluctantly said I could go. I had already talked to my older brother about this before he left home, and asked him to put in a word for me with his captain, our uncle, to give me a berth. The kedgie was almost always selected by the skipper. Owners very seldom had a say in getting men. But I also went to Stone's Cove and talked with Captain Thomas Bond, the owner of the vessel, and he gave his consent for me to go as deckhand. To my surprise, he also gave me a guarantee of $40 for the season!

I was really on top of the world now

I was really on top of the world now. I already had two cousins ten years older than I that have been captains of vessels for two or three years and, who knows?, I might also reach that lofty position some day. Anyway, I was sure of one thing. If I didn't, it wouldn't be because I wouldn't try.

The vessels arrived the first week in April; both the *Allan F. Rose*, the one my father was going in, and the *Mamie and Mona*. Both vessels first went to Little Bay East to discharge their cargo and I went lumper. My father had already sailed for the banks the day before we arrived back at Stone's Cove for supplies. We arrived early on a cold morning, and were to sail again in the afternoon. My brother and I walked from Stone's Cove to Anderson's Cove to get our clothes. We knew it wasn't going to be easy. Father had just left for the first time since they were married, and now both of us were leaving, so my brother said, "Do you think you should go?" I pondered this question all the way back to Stone's Cove. But I went on, and we sailed. All the way, going through the bay, I was thinking about my mother.

We arrived at Belleoram just before dark to take on salt and other supplies that we didn't have at Stone's Cove. As we were docking the telegraph operator passed a telegram to my uncle, Captain Green. He tore it open and it was from Uncle Charlie, the telegraph operator at my home, "You must send Arch home. His mother is heart-broken. She will never get over it if he carries on." So I was landed at Belleoram. I wanted to go so badly, but I never regretted going home. I worked with both of my uncles that summer, dividing my time between each of them.

By the end of May or early June, the vessels were arriving from their spring trips. How glad our family was to see father after being away for almost two months. Some of the vessels had a successful trip, others were not so lucky. My father's vessel was one of the unsuccessful ones. But fish or no fish, it was always a welcome sight to see the vessels return safely to port. I went lumper unloading and washing out fish in my father's place, while he spent a few days at home.

We didn't go hungry that winter

Fish was fairly plentiful in Fortune Bay that year, and I did well fishing with my two uncles. It was a good thing that I did, because my father's voyage to the banks was very poor. By the time his expenses were taken out, he had very little money coming to him. But with my help and the help of my older brother, we didn't go hungry that winter. There was still plenty of work ahead of us. When my father and brother finished the deep sea fishing voyage, the three of us fished near home using our own motor dory. When the weather wasn't suitable for fishing we went in the woods to secure our winter fuel and hunt for rabbits and partridge.

I don't think my father was very impressed with his deep sea fishing experience. He never discussed it very much. Now and then he said to me, "I wish you didn't have to go to the banks." But in his mind I guess he knew I would go, because from Terrenceville in the bottom of Fortune Bay to Cape Ray on the southwest coast practically all the men were deep sea fishermen.

Nearly every settlement had vessels

From Fortune Bay to Gaultois there were many deep sea fishermen and a large number of vessels operating at the same time. Nearly every settlement had vessels, large and small. Harbour Mille had one little vessel of about fifty tons owned by a Captain Barnes. She was named the *Thistle*, and she could outsail any other vessel around at the time.[8] She was also a nice looking craft, almost as popular then as the *Bluenose* was in her day.

There were several other small vessels operating from Harbour Mille, and Little Bay East had several, most owned and operated by the people of that place. Bay L'Argent had deep sea fishing vessels of a larger type, from sixty to seventy tons, and St. Bernard's always had a small number of banking vessels owned by Parrot Brothers. English Harbour East had their little vessel, the *Frances*, and my father's brother was its skipper. And Rencontre East had their vessels. The *Marconi* was one of them.

Many of the smaller vessels went herring seining in the spring and carried their catches to St. Pierre and Miquelon to supply the many vessels that came from France to fish on the Grand Banks.[9] When the herring seining was over they sailed to the southern Labrador for cod fishing. Belleoram, Stone's Cove, and St. Jacques schooners — the 'northsiders' — and others from St. Mary's Bay, Placentia Bay and Hermitage Bay, always did a lot of fishing on the French Shore,[10] down around St. Anthony, and on the Labrador. The northern people would begin fishing at Codroy, and work their way around Cape Anguille. There was an excellent fishing ground and a lot of fishing off Cape George and up the coast off Bonne Bay and Cow Head. Tons of halibut were caught off Cow Head by American and Nova Scotia schooners.

The northsiders would return home in June, discharge their fish and sail for southern Labrador. They'd make a straight line for Greenly Island, Forteau Bay, West St. Modeste, Blanc Sablon, and as far south as Mecatina Island. With caplin for bait, they took loads of fish in all these places, and arrived at their home ports in August, where the fish was washed out and dried.

Most of the smaller boats returned to Cape St. Mary's and

Trepassey to fish their fall trips there, while the larger vessels baited with salt squid and returned to the Labrador for their fall trips. They began fishing around the 20th of August, and tried not to be later than the last week in August. Many four- to six-dory schooners moored at Battle Harbour, where the men fished three or four miles from the mothership, while the larger vessels went up the coast as far as Indian Tickle, the end of the line in those days.

But the Grand Bank and Fortune schooners — the 'southerners' — started out in the spring just on the Western Banks — Bank Quero (or 'Banquero'),[11] Missaine (or 'Mizzen'), and down on the Grand Bank. They seldom went to the Labrador until later years. And, in the summer, every one of the five or six schooners fishing out of Stone's Cove went down on the Labrador with caplin, while the southern schooners brought it in on the Grand Banks. Later, the northsiders went to the Grand Bank just as much as the Grand Bank, Fortune, and Burin vessels. They salted all their catch in these days.

He became captain and a real 'fish dog'

Stone's Cove had quite a number of deep sea fishing vessels and quite a business was carried on there for the size of the place. The business was founded by a local man named Thomas Bond. He was an energetic young man who began his career by going to the Grand Banks. He didn't spend many years in dory before he became captain and a real "fish dog." He brought in many bumper trips of fish. I can remember people saying, "Tom Bond will soon be in with another load." He spent a few years as captain, and then began his own business. He purchased a vessel, and made another man her captain. But if the captain or one of the men took sick, he would take his place rather than have the vessel delayed.

When I reached the age of fifteen there were about twelve or fourteen vessels sailing out of Stone's Cove. Some were small fishing and freighting vessels. Thomas Bond built up his deep sea fishing fleet to six or seven ten-dory vessels, but he didn't live to enjoy what could have been a prosperous future. He had an accident on board one of his ships that shortened his life. Despite his illness, he kept his business going until he died, and, a few years before his death, he moved part of the business and his family to Little Bay East. He moved five vessels there because it was a much safer harbour. This gave the settlement quite a boost. He built a fine home there, and had a small pleasure boat. They had two children and named a vessel after them: the *Stanley and Frank*.

Mr. Bond and his wife died at an early age shortly after each other. One of his vessels, the *Flora S. Nickerson*, carried his body to Stone's Cove and while going in to anchor she ran into the cliff and broke her bowsprit. One of the crew members who had sailed to the Grand Banks with Tom Bond quipped, "It's a wonder he didn't jump out of his box." Mr. Bond's business began to wane after his death, and it eventually died out.

Pool's Cove was another settlement where there was much activity carried on by the inshore and deep sea fishery. It had between two and three hundred people, and John H. Williams carried on a fairly large business there. He and his brothers owned a large store and supplied many of the outlying settlements with food and fishing supplies, etc. And they did well with the herring fishery. They also built vessels at Pool's Cove, large and small. One of the deep sea fishing vessels, the *Loch Lomond*, carried a crew of eighteen men. A large three-mast vessel was also built there and used to export fish to European markets.

All of Fortune Bay was a hub of activity . . ..

All of Fortune Bay was a hub of activity in those days. They bring back a lot of pleasant memories, but there were also hard times. Most of the settlements had a very enjoyable Christmas season. There were teas, concerts, dancing, and mummering, for the whole twelve days of Christmas, except for Sunday. Many socials took place, and money collected was used for the war.

It was now 1916 and I was fifteen years of age. The vessels

were getting ready for the winter fishery on the southwest coast in the Rose Blanche area. Many vessels from Belleoram, Grand Bank, Burin and Placentia Bay, and other places went there. It's a hazardous undertaking and many lives were lost trying to reach the vessel or make port in blinding snow storms. While the deep sea fishermen were away, the inshore fishermen continued with the herring fishery.

Many lives would have been lost

My brother was going fishing in the *Stanley and Frank*, commanded by Captain Albert Pope from Stone's Cove. They left home about the middle of January. On a cold, dark February morning after leaving Harbour Le'Cou to go fishing on Rose Blanche Bank, eighteen to twenty miles offshore, the vessel was jogging and it was thick with snow. They were waiting for daylight to put out their dories when another vessel named the *Alma Harris*, commanded by Captain Philip Johnstone from Jacques Fontaine, Fortune Bay, collided with them. The *Alma Harris* went to the bottom. Fortunately, there was no loss of life. The men got in their dories and were picked up by the *Stanley and Frank*. My brother told us after he came home that if the two vessels had gone down, many lives would have been lost because the wind had reached gale force by daylight and they were twenty miles from the nearest land.

My father and I did well with the herring fishery, and when that was finished we fished for cod, which were plentiful that year. When there was a good fishery, everybody was busy, men, women, and children. Around the latter part of June, 1916, my father and I were fishing about seven or eight miles up in Long Harbour. We often stayed away from home two or three nights and salted our catch. When we returned home for supplies we brought a load of fish with us.

On this occasion we were only away one night when (for some reason) father said, "We'll go home tomorrow morning after we haul our gear." This we did and when we came near home we noticed all the blinds were drawn in every house, which was the custom when a person died. When we arrived at the wharf we saw mother coming to meet us and she broke the sad news. The same man that took my father in dory with him on the Grand Banks the year before, died on board the *Warren M. Culp*, from Stone's Cove, commanded by Captain Reuben Elms.

When the spring trip was finished and their fish discharged, the *Warren M. Culp* took its first caplin baiting and sailed for the Grand Banks. After reaching the banks, Captain Elms chose a position where he hoped there would be lots of fish. The vessel anchored and the gear was set. The fishing was good, and quite a catch was brought on board that day. After supper everybody began dressing the fish. Around midnight this man had a few minutes break. He picked up a shellfish from the deck and went to the cabin where a fire was burning in a hard coal stove. He put the shellfish on the burning coals and cooked it for a few minutes, then ate it. Within hours he died.

The trip was abandoned, and the man was brought to Stone's Cove for burial, a distance of approximately three hundred miles. We had no long-range means of communication then, no wireless or ship-to-shore radio. The vessel came in with her flag at half mast. What anxious moments it must have been for those people in Stone's Cove, watching one of their vessels coming home with their flag half mast, and not knowing whose brother or husband the victim was! This sad event cast a gloom over Anderson's Cove as the man left a wife and four small children.

He told me I didn't know what I was getting myself into

The year 1916 was a good one for Fortune Bay. There was a good herring fishery in the winter and early spring. Cod was also plentiful, and prices were high. The war accounted for this. My brother had a fairly good year deep sea fishing, and I was anxious to get into the game. But my father didn't feel the same about it, and he told me I didn't know what I was getting myself into, especially at an early age. My brother arrived home around the middle of October, and we began fishing together. Father stayed at home and repaired the nets, etc. We went on the jigging ground before daylight, and got all the necessary squid for bait. My brother was skipper, and he said, "We're going to fish exactly as we do on the Grand Banks." I was very glad to hear that. He taught me how to bait any amount of gear. After all, he was almost eighteen years of age and had completed his third year of deep sea fishing on the Grand Banks.

My father was a very eager man

My father was a very eager man, and he was known as one of the best fishermen in our vicinity. I can see the smile on his face now when we'd arrive from the fishing ground with almost a full dory load of fish. He was always there to meet us. By the end of 1916 our year's work brought in altogether a little more than $1,200, and that was a lot of money then. Times had not been very prosperous until then. There were ten of us in the family, and we took all we made in food, clothing, repairs to our home, etc. I got my first bought suit of clothes that year. Mothers usually made all the clothing for the family, especially in the little fishing settlements around the coast.

The year 1917 began as usual, with celebrations, concerts, dancing, parties, etc. People came from other places to go hunting in Long Harbour country for caribou and other wild meat. My brother decided not to go winter fishing this year. Instead he was going fishing in one of the Grand Bank vessels where some of our cousins had been fishing for the past few years. I was now sixteen years of age and had been trying to get my parents' consent to go to the banks this season. I wanted to begin by going in dory, but it was difficult to get a man to take a sixteen year old boy in dory, especially if the boy was not big and strong looking.

I was determined to go to the banks too

By the end of February all the men from Anderson's Cove and Stone's Cove had their fishing berths, and I was determined to go to the banks too. I decided that, if I couldn't get someone to take me in dory, I would try hard to get a chance as deckhand.

Father now had a motor dory and I could not understand why my younger brother couldn't take my place with him. When we were fishing we had to row to the fishing grounds. I was determined to get away somewhere, and what I really wanted was to get in one of those beautiful vessels that I watched coming and going.

But my father paid a price for me to stay at home with him another year. I was very fond of accordion music, the music played at dances. I often wished I had one, so that I could learn to play for dances sometime, but I knew it was impossible. In the first place, they were very expensive, and a very scarce item because of the war. But I knew where there was one, because I saw it when I used to carry the mail with my uncle. It was at Femme, a little place about five or six miles by water from my home. It was a large accordion, and I admired it. My parents often heard me speak of it. One day when I was talking about going on the vessels, my father said, "Would you like to have that accordion that you've been talking so much about?" and I said, "I sure would." "Well," he said, "if you will promise your mother and me that you won't talk anymore about going to the banks this summer, I'll get you one." I got another boy with me and we went to Femme to see if this man still had the accordion. It was still there, but what about the price? The man said, "You can't give me enough money to buy this machine." $10 or $12 was a lot of money, and the accordion wasn't new. I offered $15, but he wouldn't sell. Finally, after hesitating awhile, he said I could have it for $20. I went home and told my father. I knew it was a big sacrifice for him to make, but it wasn't too long before I had the accordion.

Practically all the crews that manned the Grand Bank vessels were from Fortune Bay

My brother sailed this year in the *Bella Bina P. Domingos*, a Portuguese fishing vessel at one time, but now owned by Samuel Harris and Company of Grand Bank. This company owned quite a few large fishing vessels and some three mast vessels that carried fish to Portugal and other European countries. Other firms in Grand Bank also owned deep sea fishing vessels. There were few real Grand Bank dory fishermen there. Practically all the crews that manned the Grand Bank vessels were from Fortune Bay. One or two vessels would go all around the bay and take on board all the men that were going to the Grand Bank vessels.

There were fifteen or sixteen men from Anderson's Cove alone, and when the men left it was a general clear out, and a lonely time for those who were left behind. There were only two or three of my age left. I had been practicing on the old accordion and making fair progress. But when the men went away to join their fishing vessels, it wasn't the accordion that was uppermost in my mind. But I kept my promise and continued fishing with my father. There was plenty of work as usual, and my younger brother (Stanley) was a great help to us when school closed. The war was still raging, and it didn't look good at all. My uncle was the telegraph operator, and every afternoon he posted the war news on the bulletin board outside the office for everyone to read.

When Long Harbour became free of ice, the spring fishery began again in earnest. The herring and cod were very plentiful at this time, and we brought in many dory loads of fish. But we were isolated up there in Long Harbour. It was almost like being in jail. I think my father saw I was getting restless, so he said one day, "I think we'll take in all our gear, leave for home, and get ready for the caplin trip." In May the vessels arrived again from the spring trip. They had done well. They would be in Grand Bank for about a week discharging their fish, and then leave for the caplin trip.

A few days after my brother's return we received a

telegram from him in Grand Bank, saying he had joined the Navy. It was a sad day when that news came, almost like a funeral. Everyone thought, especially in a small settlement, if you went to war, that was the end. And for many a mother's son, it was the end. Telegrams were sent asking him to change his mind, but to no avail. After awhile, I think, they gave him credit for going. He spent some time training in St. John's, and in August he and another young man from Anderson's Cove came home on leave. They were about to return when they received a telegram saying, "Wait for further orders," and they didn't have to go until December. We all went fishing together while he was home.

It wasn't easy to get a doctor in those days

A few weeks after my brother came home on leave I was taken ill with a bad sore throat, and within days my sister and youngest brother were also ill. It wasn't easy to get a doctor in those days, but a clergyman who had some medical training was stationed at Pool's Cove, and he was called in.[12] He said it was "a bad sore throat." We were running a high temperature, and it was serious. My sister and I were close to death and, later, the doctor said it was a wonder we didn't die. After a while another two or three people became ill with the same disease, so they called another doctor who was stationed at St. Jacques. He didn't pronounce it diphtheria, but some of the older people in the settlement thought it was.[13] Finally, after a couple of months, we were well and no one else caught the disease, not even mother who attended us so well. Just before my brother got his orders to leave home he mentioned something about men with sore throats on their training ship in St. John's. But, now that we were well, no one took that very serious.

The year 1918 came in as usual with the ringing of bells and horns blowing from the vessels in the harbour, and other festivities. In the early part of January I was taken ill again with this dreaded disease. Other people became ill, and one young man in Stone's Cove died. The doctor from St. Jacques

I Never Gave Up Once in My Life

Young Arch, at about seventeen years of age, following his first voyage aboard the *Bessie Macdonald*, c. 1918.

was called in again and this time he pronounced it diphtheria. All homes were quarantined. It looked serious for everybody, but especially for me, as it was the second time for me in less than six months. We were fortunate that no one else in our family caught it this time. Every family in the place was ordered to fumigate their home with everything in it, and the fumigating went on for three days and three nights, then all the washing began. It was a tremendous amount of work and the weather was never colder. Finally, the disease was beaten.

Counting fish, we got to have the best men

It was February now and I was seventeen years old, and my contract with the accordion had run out. I wasn't very big then, and I'm not big now. I never weighed more than about 132 pounds; a small child. But I was looking around now for a man to take me in dory with him. I decided to ask Mr. (George) Fizzard, one of the experienced Andersons' Cove fishermen. I went to him and offered him $20 to take me.[14] "No," he said, "No, I wouldn't think of it, Arch." He cut it right short, and I wanted to get to the banks so bad. Well, I went back to our house. I went in disgusted, but my mother and father were glad because they didn't want me to go to the banks as early as that.

After a little while I went to him again, and offered him $40. Quite a bit of money then. I said to him, "As soon as the voyage is over, I'll put the money in your hand." He said, "No. My son, do you realize the hard work and hardships you got to go through out there?" "Yes, sir," I said, "I realize it." "Well, you haven't realized it much," he said, "because you're not old enough to have time to realize it. You got to work when you goes out there! Counting fish, we got to have the best men." He was a top dory fisherman wherever you went. He said I was too small and not strong enough to do the work expected of me. No luck. No luck.

I went in the house again, and my mother and father didn't have to ask me the results. They could tell by the look on my face that I was disappointed. After a day or so I said to my father, "I'm going to ask once more, and I'm going to offer him $60." That was a lot of money in those days, and my father said, "Son, I hope you know what you are doing."

It was getting up in winter by now, close to the time to get ready for the spring fishing. I went to him the third time and I offered him the $60. I said, "I'll do everything I can, I'll work so hard as I can." "Gosh," he said, "that's a lot of money to turn down. Well, seeing you're so determined, I'll take you on, and we'll give it a try. I haven't decided where I'm going next year yet, but I've had several telegrams from different skippers in Grand Bank asking me to go with them. I had one there a few days ago from Captain Harry Lee, and I think I'll wire him and tell him that I'll go with him. We'll be sailing from Grand Bank in the *Bessie MacDonald*."[15]

Chapter Three

Voyage to the Grand Banks

Be careful what you wish for. It may come true. One morning in late March 1918, together with Mr. Fizzard, his patron, their clothes bags in hand, Arch walked the path to Stone's Cove to enter into the higher mysteries of banks dory fishing. He was a small and quiet, seventeen-year-old. His capability, mettle, and potential were not obvious.

He was like thousands of other young men of his era raised in Newfoundland outports. His formal education was limited, but he grew up living and working beside hard-working men and women engaged in inshore and onshore economic activities. He was armed with a self-confidence based upon the acquired knowledge, skills, and values that enabled his elders to make their living. He knew about inshore fishing, small boat seamanship, and subsistence activities that balanced returns from fish production traded with merchants. And he already knew much about banks fishing's technical kit, work roles and hazards, operational regime — the baiting cycle, how they fished, processed and stored their catch, and discharged it in port. Urban young men of the same age would be hard put to match his ability to "stand on his own feet".

But he would need more than these resources and his pact with Mr. Fizzard, to become a respected banks dory fisherman. Banks fishing was industrial work with a vengeance. It would take gritty determination and an eagerness to succeed that would be severely tested. He must first get his "sea legs", what many view as the seafarer's first test. And, where every fish counted, he would have to meet work demands that strained every fibre in his body. Having passed these tests his further ascent up the ladder of banks fisher authority required years of hard, broadening experience. And it would hinge upon the encouragement and judgement of kin and others, and good luck.

Although experienced dory fishermen, like Mr. Fizzard, might sponsor a first-timer, each fisherman had to "look out for himself". He served at the absolute discretion of the vessel captain and her owner. And unless he was the vessel's owner, the captain served solely at the owner's discretion. Owner and skipper expectations, orders, and judgements were supreme. The aspiring banks dory fisherman had to establish a reputation for reliability, effectiveness and obedience. The system of credit dependency that dominated the fishery channeled conformity. The chance to sail in dory and make a living hung in the balance.

Once in dory, individual work performance was driven by the skipper's use of "the count", an incentive scheme where an individual's earnings depended upon his dory's catch as recorded by literally counting every fish heaved aboard. Hoping to insure their success against poor fishing days, dory crews felt compelled to use practices or 'stunts' that increased risks already inherent in banks fishing. Countless dory crews went astray, some never to be seen again. Old dory hands remember this era ambivalently. They recall intense labour, the achievement of being "high dory", dispair over failure, and lives lost, and men who "went to their graves cursing the count". And some still ask: Was the count necessary to make men work?

1918-1927

I left Anderson's Cove in the last week of March 1918 and carried my clothes bag to Stone's Cove, where I joined the vessel *Dauntless*, commanded by Captain Arthur Scott of Little Bay East. She was owned by Samuel Harris Limited, the same company that owned the *Bessie MacDonald*. There were seventy or eighty men aboard, all being taken out to join this company's vessels. We arrived at Grand Bank[1] on a Saturday night after a fine passage. Some of the men said to me that day, "Young fellow, don't expect smooth sailing like this all the time. You will see seas as big as mountains before this voyage is over." It was my first time in Grand Bank, and on Sunday I went ashore with my dorymate for some sightseeing. There was quite a difference between Grand Bank, with its 3,000 population, and Anderson's Cove with only 100. Mr. Fizzard said, "We'll go down to the barracks tonight." That was the Salvation Army. That night it was dark as pitch, no lights, going up through town. I took hold of my dorymate's coattail, just like a stray, so I wouldn't lose him.

The Salvation Army Hall was filled to capacity when we reached it, and there was a lot of excitement. Bands played up on the platform, and men and women stood up to give testimony and thank God for his gifts. Whenever we were in Grand Bank after that for a few days, we would go up to the Hall every second night and on Sundays, and just stroll from one end of town to the other.

The *Dauntless* had just completed her winter fishing trip in the Rose Blanche area. During the last week in March and the first of April the vessels that didn't go winter fishing were being made ready. Monday morning was exciting for me. It was my first time working on a vessel and preparing for the bank fishery. There was a lot of work to be done and many questions asked before the *Bessie MacDonald* was ready to sail.[2] It took about a week to get the vessel ready. The men had to get their trawl gear all hooked and lined up, and they had to get their dories ready and well fitted to make them seaworthy.

Each dory would have forty lines of gear, each line forty-five fathoms long, with approximately seventy-five seds[3] on it, and a fish hook on each sed. Each man in a dory is responsible for putting half the gear in order, which meant he had twenty lines of gear to get ready. We had many sore and sometimes infected fingers before the work was finished. The larger lines had to be put in order with a small, sharp point splicer that often struck the finger and caused the soreness. The cook was a busy man too. He had to cook for twenty or twenty-five men. It would be the wee hours of the morning before any of us got to bed.

I smelled a rat... What's this I hear about a guarantee?

We were aboard, getting our vessel ready on the second day when I heard the men talking about wages. I smelled a rat, so I asked one fellow, "What's this I hear about a guarantee?" He said, "Look, me son, this skipper is getting along in years and he haven't been gettin' much fish lately, so the firm had to give wages in order to get a crew to go with him. Didn't you know dat? Ain't you got wages? All we fellers got $600 wages, and if we makes more we'll get more." When I asked my dorymate about it, he didn't say much. "Oh, yes," he said, "I've got wages, but you're a young feller, so I don't expect you to get wages. Anyway, we'll make as much as anybody, so don't worry."

I didn't say any more about it then, but I was still curious, and I said to myself, "If you don't help yourself, no one else is likely to help you." So, on the third day, I sneaked off from my work and went up to the manager and owner's office. When I knocked on the door, a gruff voice said, "Come in." I went in and he looked me up and down and said, "What do you want, young feller? What can I do for you?"

I said, "Sir, I'm on a vessel down there in the gut now, gettin' ready to go to the banks, and I understand the men got wages. And I was wondering, sir, if I could get wages too?" Then he asked me for my name and age, and what vessel I was in. I told him and he said, "Oh, seventeen. Small man. I don't think you've been to the banks before." "No, sir, this is my first year," I said. He asked me who my dorymate was and when I told him he said, "Oh, Fizzard. He was with us last year and he's a good man, a real highliner. Well, I'll tell you what we'll do. I'll give you $300. You should make that much with Fizzard, and if you banks anymore, of course, you'll get the value of it like the rest of 'em. The final settlement, you'll get it just before Christmas."[4] I accepted with thanks and went back down to our vessel. I went aboard and when I went up to my dorymate and told him that I got wages and shares too, he was really surprised.

In those days we counted fish. Any two men in a dory who caught the most fish would make the most money for the season. Before you began fishing, the captain would give orders about how many lines each dory would use. But if any two men were smart enough and didn't mind work and kept up with the rest of the crew, they would add a few extra lines. And this was often done. The skipper had a tally board with each dory's number on it, and a line drawn down through the board between the numbers. I've been told this was copied from the French fishermen.

Grand Bank, March 26th, 1918

It is hereby agreed between Arch Thornhill and Samuel Harris that he shall serve the said Samuel Harris on order, from the 26th March 1918 until the end of the fishing voyage in the capacity of fisherman or anything else in his power for the good of the voyage or his master's interest as he shall from time to time be ordered; and it is also agreed that the said Arch Thornhill shall not during the continuance of this agreement drink any intoxicating liquor under a penalty of Twenty Dollars for each and every time he may be convicted of such an offence; and in consideration of his services being in all respects well and duly performed, without any hindrance or neglect, according to the custom of the fishery, he is to have as wages: $300 and share out of his own dory.

    Witness: (unsigned)

    (Signed):   Samuel Harris

Lives were lost by overloading the dories

When the trip was finished, the skipper added up the tallies and put the board in a conspicuous place so that everybody could see for himself. Some men kept their own tally book and compared it with the skipper's. After the vessel arrived at its home port, the captain would take the tally board to the company's head office. When the fish was discharged, washed, dried and weighed, the count of fish was made up per quintal. When this was done you knew how much one fish or 1,000 fish would be worth, depending on the price paid. For example, if a dory (two men) had a count of 5,000 fish when the trip was finished and it fetched $20 per 1,000, each man for that dory would share $100 for his trip. But there was much controversy and bitterness about counting fish, and lives were lost by overloading the dories.

So it remained to be seen whether I would just make my wages or add to it. One thing I did know: there was a long season ahead and plenty of work. By the end of the week we were about ready to sail. The vessel was all fitted out, the dories all nestled in one another, the large eight inch cable on deck, and the 800 pound anchor with a large wooden stock in it on the bow. Salt and food supplies were on board, and the men had their gear in their four trawl tubs, their dories numbered from one to eight, and everything in each dory numbered.

Monday, April 1, 1918 was the big day, my first day sailing for the deep sea fishing banks. First, we went across Fortune Bay to Connaigre Bay for a baiting of herring. We took eight dory loads at $20 per dory load. The herring was put in bins in the hold of the vessel and covered with enough ice to keep it for about three weeks, and all the water tanks were filled for the three week trip. We stopped at Grand Bank on our way along, because there was almost always something forgotten.

Coming across the bay to Grand Bank, I had my first taste of seasickness. The wind was strong from the southwest, so the captain decided to stay at Grand Bank until morning. The wind was still strong when we left the next morning, and it was dead ahead, which meant we had to beat to windward all day. The sea was pretty rough. To make matters worse, it was our third watch. The rule was there were two men on watch at a time, and each man would stay at the wheel steering the vessel along at half hour intervals, while the other man was on the lookout. Later on this rule was changed from two men to three. By this time I was really seasick. I was washing the decks and everything. I stuck to my half hour steering interval, but it wasn't easy, especially when the

Fishing and trading schooners line wharf at Grand Bank, some time prior to 1914. Flag-decorated schooner at far right is the *N. Fabricius*, loading codfish for transport to distant markets. Based at Marstal, Denmark, she was the smallest Danish schooner in the Newfoundland trade.

skipper came along and said, "What in the hell is wrong with you, young fellow? Can't you steer straighter than that?" I wondered many times why he didn't throw me overboard.

The rule was to touch in St. Pierre going along so that the men could get their tobacoo, cigarettes and, if any of them were lucky enough to have a few dollars, rubber boots and oil clothes, because they could get it cheaper there — duty free. You wouldn't have to pay half as much for it as you would in Grand Bank or Stone's Cove and other places in Newfoundland. And the sugar was cheap out there, so you'd smuggle the scattered bag. Coming back from the banks, as a rule you wouldn't touch in there. It depended on whether or not the skipper wanted a drop.[5] If he had an excuse to go ashore, sometimes you wouldn't even anchor, just heave up and lay around there in the vessel until he came back.

We arrived at St. Pierre just about dark, and what a day for the first day out. There were others seasick on board, but most of them would get over it quickly. And the sea wouldn't roll down big enough to take any effect on some of them. But poor me! Would I ever get over it? This was my first time in St. Pierre and the first time I heard people speaking French. The port was filled with big, three mast, square-rigged vessels. Most of them came from France to fish on the Grand Banks and used St. Pierre as their headquarters. There were also a number of our own vessels there.

I couldn't care if someone threw me overboard

The following morning we left St. Pierre bound for Banquero and Mizzen Bank. As the day passed I lost sight of land for the first time. I was feeling miserable. The vessel was putting her bowsprit in the water continually. Towards night sail had to be shortened, and I couldn't care if someone threw me overboard. My second watch came on just after dark. I stuck to the wheel for my two half-hour intervals. The skipper was still on my back for not steering straight and I wondered if the old bugger was ever seasick in his life.

We arrived at our destination just before dark and anchored. I hadn't eaten since leaving St. Pierre, and the way I felt now, I wasn't sure I would ever eat again. After anchoring with approximately 200 fathoms of cable and our 800 pound anchor, an anchor watch was settled; one man keeping watch for an hour. Later, that was reduced to half-hour anchor watches. About ten o'clock that night I was called out for my first anchor watch. It was very cold, and dark and lonely until you got used to it. The man that I was relieving gave me the orders to pass through the watch.

One-thirty in the morning, I was asleep in berth when the watch came down and that God-forsaken call came that I'll never forget as long as I live: "Heave out, boys! Heave out and bait up." I opened my eyes and looked out — What's he saying, I wonder? I never heard that call before. I got up to the cabin to bait up our gear, and what a miserable morning it was, snowing and freezing cold. One man from each of dories number one, two and three would get in the hold of the vessel and take the bait out of the bins. The next time the gear would be baited, the other three men from these numbers would take their turn getting in the hold. It would go on this way until every man had his turn. Each man had two tubs of gear to bait with ten lines in a tub. The bait was covered in ice and you had to dig it away with your hands, pick it out, and put it on your hooks.

It was the first time I ever baited on the banks, and here I was, seasick. After baiting our gear for an hour or so, the cook called out, "Breakfast." I was too miserable to even think about eating. My stomach was beginning to feel like a piece of raw beef. I managed to bait one tub of gear while my dorymate baited two of his and one of mine. I can imagine what was going through his mind. The rest went down to get their breakfast, but I couldn't go and I went in the dory without my breakfast.

I began to pray I would die

When daylight came everybody was ready to leave the mothership and set their gear. Each dory had their own course, and when all the gear was set, everybody returned to the mothership to get a snack before hauling the gear back again. I took some hardtack with me. I thought I might nibble a bit, but my stomach wouldn't take it. I couldn't even enter the forecastle without vomiting; that's the worst place on the vessel for seasickness. I began to pray I would die, and even coaxed the men to throw me overboard. When the time came, we took back our gear. Fishing was fair. Most dories had a couple of loads, and late in the evening, when all the dories were on board, the old skipper was pleased with the first day's work. Around midnight all the fish was dressed and salted in the hold. Another anchor watch was settled until two o'clock, when we would heave out again and go through the same routine.

The first three days we were anchored it was very good fishing, also plenty of hard work. I worked hard. I had to because I was small. But I had to do the work they'd give a bigger man. Everyone used the same amount of gear. There was a strong current running and we were fishing in deep water, which made it much harder. My dorymate wasn't getting too much help from me. I don't believe I ate 10 ounces of food since leaving St. Pierre. Several times I got weak. I worked so hard, sometimes I saw stars and passed out up by the cabin when baiting up the gear. And when I went down to my breakfast, as quick as I sat down I passed out. My hands were so cold, and I was tired. But I never, never once blacked out in the dory when haulin', doing hard work.

The third day, as we were coming on board late in the evening, my dorymate said, "You will have to leave the vessel when we get home. The skipper will be lucky if he don't have to abandon the trip and carry you in. It's too hard on me as well as yourself." It almost killed me when he said these words: "You will have to leave the ship."

But I remembered one of my cousins saying to me before I began fishing, "There was never anyone yet who died of seasickness. If there was, I'd be dead."[6] From that time on I began to force myself to eat a little. We came alongside the schooner and discharged our fish, and I went to the foc'sle where the cook had some nice looking buns laying on the counter. I tried one, but I couldn't keep it down. I tried another, and it came back. But the third time I kept it down and felt much better. The fourth day, when we left the ship, I took a few buns with me and from then on I felt a little better each day.

My dorymate could tell the difference in me right away. "Boy, you're bucking up, Arch, perking up a bit!" I still got seasick, but I began to eat and keep it down. I think the message I got from my dorymate did the trick. Another thing that helped me was, I didn't get sick in the dory. It was only when I was on the vessel.

I was determined to get there and fought it off

We went through the same routine every day when it was suitable to fish, and before the trip ended my dorymate didn't have to bait any of my gear. This first trip lasted almost three weeks, and I knew my parents were anxiously awaiting a telegram from me and I was anxious to hear from them. I knew by this time that I would be able to carry on, but I also knew it would be a long time before I'd get over seasickness. Year after year I suffered like that. But I was determined to get there, and fought it off. And some people don't know what it is to be seasick.

We arrived at Grand Bank with a successful baiting, and we were second to high dory in the fish count. My dorymate was pleased. I couldn't get up to the telegraph office quick enough to wire my mother, "I don't get seasick. Arrived in from the banks with 450 quintals." I could steer the vessel much better, but there was still room for improvement. The old skipper acted more kindly towards me and seemed very pleased every time we came alongside with a doryload of fish. And when we arrived at Grand Bank he invited me to his house on Sunday evening. He had a nice looking daughter who played the organ, and the two of us had a sing-song while the skipper and my dorymate were talking out in the sitting room. On Monday we did not discharge our catch but took on supplies, and on Tuesday morning we left for the Grand Banks.

It was May now and the weather was getting warmer. It was much better fishing on the Grand Banks than on Banquero. The water is much shallower, there's less current running, and we used little more than half the gear used on our previous trip. We set our gear and leave it in the water overnight. We would never leave the gear out overnight on Banquero. The weather was perfect on this baiting; scarcely any fog at all on most days. Some of the older men said to me, "Boy, by the time you punches in as many years as I have you'll have it a whole lot worse than this."

It was almost as if we were in a field of woods

One day when the fog cleared it was almost as if we were in a field of woods. There were vessels in every direction. Newfoundland, Nova Scotia and American vessels, also large three mast square riggers from France. The particular place where we were anchored was good for cod, and that accounted for so many vessels being in that vicinity.

I enjoyed my work much better than I did on my first trip. I still got sick on board the vessel, but it didn't hurt as much and I ate more. Fishing was fair and my dorymate and I were keeping up with the highliners. We arrived back in Grand Bank on May 24th with our first spring trip. We were still second in line with the high dory. We had eight hundred quintals for our trip. One hundred and twelve pounds for a quintal when it's washed and dried. It's a fair trip, but some vessels had more.

Grand Bank was a very busy place when the vessels arrived from their spring trips and discharged their catches. There were twenty-five or thirty banks schooners there alone, each with from fifteen to twenty-five men — 400 or 500 fishermen. Besides the fishing vessels, there were three mast vessels that took the fish to European markets, and shipyards building vessels for use in the foreign trade. One could almost walk from one end of Grand Bank harbour to the other on vessels large and small. It was a beautiful sight. And there were many shore fishermen in Grand Bank, using cod traps and trawls, lobster pots, and salmon nets.

And Grand Bank never had more than one policeman all the years I was going there. There was always a drop around because you always touched in at St. Pierre going out or coming back. But I never heard of any fellows getting disorderly when they went ashore at night. And we had mixed up crews. There'd be Catholics and Protestants, but never any disturbance over religion. The voyage would probably be half over before I might find out that a fellow was Catholic, and I don't know that he ever found out what I was.

When we were in Grand Bank discharging and washing out our fish, you'd be tired when you finished work, but there were few nights when you didn't go ashore. There was little entertainment to go to, only the Salvation Army and Methodist Church, and there'd be a small shop, like an ice cream store with a crowd inside. But no dancing then. They believed then that you'd go straight to hell if you danced.[7]

After about a week in Grand Bank, discharging, washing out fish, and taking on supplies, etc., we secured a baiting of 10 or 12 dory loads of caplin from up along Grand Bank, and sailed again for the Grand Banks the first week in June. The war was still raging and there were rumours of German submarines prowling the American coast and the Grand Banks.

I'd been told about the fog on the Grand Banks at this time of the year and we sure had plenty on this trip. The caplin trip was the shortest of the year, lasting about one month. The small caplin fish used for bait doesn't keep as long as the herring bait, which means each baiting is short; about five or six days if the weather is good and the sea not too rough. After five days we had our first baiting fished. The first caplin baiting was secured on the southwest coast, Fortune Bay or Placentia Bay. Then the vessels went in to Aquaforte, Ferryland, Cape Broyle and Calvert on the Southern Shore to get their caplin baitings for the remainder of the trip.

There was always great activity in the Southern Shore communities. As many as forty or fifty vessels would be there at a time. The men from these places secured ice and put it in large ice houses during the winter, and the shore fishermen had their own caplin seines. The Newfoundland fishing vessels had their own seines too, but the Nova Scotia vessels bought their caplin bait from the shore fishermen. This created business in these places, and all the ice houses were empty by the time the caplin trip was finished.[8]

The vessels had no engines in these days, and it often happened, if the weather was calm, that they were so long getting to the fishing grounds that the bait in the hold was spoiled. You had to dump it and return to land again for fresh bait.

We finished our caplin trip around the 20th of July and landed our catch at Grand Bank. Then we went over to St. Pierre to put the vessel on dock. And that's where the squid — our next baiting — struck in first. So the skipper bought our bait from the Frenchmen.

We left one afternoon in the first week of August and sailed for the banks again on our fall trip with a squid baiting. We heard rumours before we left that some Nova Scotia fishing vessels had been sunk on the Grand Banks by enemy submarines, and sailing out of sight of land that first day, just before dark, we heard distant rumblings like gunfire.[9]

The weather was fine and clear, and the skipper and crew seemed a bit nervous but we sailed on. A few hours after dark we saw this ship steaming toward us with a light making quick flashes. She came alongside and threw a powerful searchlight on us, and everybody rushed out on deck with oilskins and rubber boots. It was a British cruiser. She came close by and ordered us to put out our navigation lights and make for land as quickly as possible. A small steamer, an old whaler[10] bound for Sydney for a load of coal, had been sunk in the vicinity just before dark. We left immediately for St. Mary's Bay, where we fished the rest of the fall.

We came to land for a few days and there we heard the news about that steamer being sunk and that its engineer was killed. When the steamer's crew was taken on board the enemy submarine they found a number of other Nova Scotia fishermen already there. The submarine sailed as far west as St. Pierre Bank, where it came across the *Wally G.*, another Newfoundland fishing vessel, commanded by Captain Wilbert Moulton of Grand Bank. Captain Moulton was ordered to put his dories out and take all the men from the submarine, put them on board his vessel and get to land without delay. The captain of this enemy submarine was more humane than others. When some Nova Scotia vessels were sunk the men had to row their dories 100 miles or more to make land.[11]

All the vessels left the Grand Banks then and many abandoned the voyage for the season. We continued to fish around Cape St. Mary's until the end of September. There were 50 or 60 jackboats — two to four dory schooners — fishing off St. Mary's Bay that year. I often saw them anchored in places, jigging squid. There were hundreds of those jackboats around in Placentia Bay, St. Mary's Bay, and Hermitage Bay, and some in Fortune Bay.

The happiest time of the voyage was when the skipper said, "Take off your hooks"

It was very stormy the last day we fished, and we wondered if we would get our gear and make it back on board safely. It was pretty rough by the time the last dory got aboard. The happiest time of the voyage was when the skipper said, "Take off your hooks." Then you knew that's the last day of the voyage. If you were on the Labrador, on the banks, anywhere, you'd wonder when those few words would come. And we weren't long getting our gear up out of the cabin and taking off the hooks. We shaped our course for home and arrived at Grand Bank the following day and my first deep sea fishing voyage ended.

The first thing we did when we arrived was discharge and wash our fish and put it on the beaches. Then we gave the vessel a thorough cleaning, took all fish hooks off the trawl, put all the fishing gear, trawl buoys, anchors, the 300 fathoms of cable and all reserved gear in a warehouse. If the vessel was to be moored in Grand Bank harbour, all sails were dried and put ashore in the sail loft. Finally, each dory crew took their dory ashore on the beach somewhere and scrubbed her out with sand. Then they were all carried up and nestled in each other by the owner's store. When all this was finished, "Jack was as good as his Master." He's his own boss. When that last thing is done, the skipper can't come down and give you orders. He might ask you to do something in a kind way, but he won't tell you to do it.

Some didn't earn their wages

The following day was very exciting. It was "settle up" time. We didn't have a big voyage, but my dorymate and I ended up highliner — high dory. Some didn't earn their wages. They were low dories because they didn't catch any fish. We worked hard, but my dorymate deserves most of the credit. I'll never forget how I suffered from seasickness. I learned a lot, but there was much more to learn, and I looked forward to going to the banks again next season. We had only a partial settle up because the fish must be dried and weighed before we got a final settlement. But I knew I was going to make more than my wages. The crew settled on their $600 wages and hoped to get a few dollars more when the final settlement was made around Christmas.[12]

I was a very happy boy when I came out of the manager's office. He wanted to know if I had any idea how many fish my dorymate and I had caught for the summer and I said, "Yes, sir, we keep tally of our fish in a small book." He said, "I'm very pleased with the report I've received about you, and don't be embarrassed when I tell you I know how seasick you were and how you fought to overcome it. I'm going to give you $500 now and, the way things are looking, you should have another $100 coming to you around Christmas time. And you will always get a berth on one of our vessels if you need it." I thanked him and left in a very happy mood. I went straight to my dorymate back aboard the schooner, and put the $60 I had promised him right in his hand.

When you settle up, every man would buy his winter supplies from the company he was sailing from. This company alone had about fifteen vessels, and 80 or 90 per cent of the men that went in their vessels were from Fortune Bay. Sometimes the company's vessels took the men and supplies to their homes, but this year my dorymate and an older cousin purchased a small Grand Bank fishing schooner, and the Anderson's Cove men put their supplies on board this little vessel. I arrived home with them from my first fishing voyage, the first week in October 1918.

I think I had as much in supplies, clothes etc., for my brothers and sisters as any man on board. And Jumbo tobacco for father, dishes for mother, and shoes, dresses, etc. for the younger members of the family. There was lots of excitement, and I put the remainder of my money in my father's hand. It was a happy time and everybody was home safe and well. But I wasn't to be well very long. 1918 was the year of the dreadful flu — Spanish Influenza. And there was a new school teacher at home that year. She became a great friend of my cousin who was then the telegraph operator. I saw her for the first time the day we returned from the banks. I was down in the schooner's hold, passing the freight up, when some of the men called out that she was going by, "There's the teacher." She had come over from her home in Pool's Cove. We met later and in a couple of weeks we started walking around together.

Doctors weren't very near or very plentiful around our way

One morning shortly after I returned home I didn't feel too well and I began to get worse. Within a few days I was delirious and so weak I couldn't stand. Two of my cousins also became sick shortly after I did, but they weren't as badly stricken as I was. Doctors weren't very near or very plentiful around our way. The minister at Pool's Cove, however, was also a medical man, and he was called in. He ordered the school closed and told mother it was a wonder that I didn't drop dead. He told me to stay in bed and he would be back again within a week. I had almost died from diptheria little more than a year before. He came back within the week and said he felt sure I was out of danger, but to rest in bed awhile longer. He gave me medicine and I gradually got my strength back, but it was thirty days from the time I took sick until I got out of doors again. And the war was over the day I got out (November 11).

It was early in December before I began to get my strength back. The flu still raged in different parts of the world. The teacher returned from Pool's Cove and school opened again. Her brother's wife died from the flu while she was at home. I was not well enough yet to work with my father and brothers at the herring, but I looked forward to joining them early in the New Year.

It was a year I'll never forget

A few days before Christmas I received my final settlement from Grand Bank for the season's work. My dorymate and I made $650! I received a cheque for $150, which meant I made $350 over my wages. It was a year I'll never forget. I worked probably as hard as I ever would, sometimes with the odds against me, but I was glad I was able to stick it out, learn many things, and, above all, reassure myself that nobody would ever die of seasickness. A lot of blood came up from my insides that first year I went to the banks, and a good many years after. But I didn't mind it too much afterwards, because I knew I could eat a bit and strenghten up.

The war was over and relatives anxiously awaited the return of their loved ones. School closed for the Christmas holidays, the usual festivities took place, and 1919 came in as usual. But the flu was still prevalent in many places. Pool's Cove was one of these places, which meant the teacher was delayed and school was closed at Anderson's Cove. But February month was happy for us. My brother and another young man returned from the war, and they were given a reception that lasted until the wee hours of the morning.

As the winter passed I began to inquire for another fishing berth. It didn't take me as long to secure one as it did the first time. I secured a berth with one of the "high line" skippers in the *Carrie and Nellie*, an eleven-dory schooner, and my dorymate was my uncle Wilson Green from Little Bay East. My brother was going on the same vessel and I was glad, because he had more experience than I, and would be able to show me many things, because there was still much to learn. We left home the last of March to join the vessel at Grand Bank. Our spring trip was a good one. We loaded our vessel on the last baiting and brought some fish home on deck. But we had some rough days, fishing. Captain Alex Smith was in charge — a fish dog, one of the big fish captains. He was about 30 years of age. One day on the Grand Banks while anchored and too stormy to fish — and, believe me, when Captain Smith kept his dories on deck, it wasn't calm — we saw a vessel anchored about four or five miles from us. Every now and then we could see the dories going alongside and leaving the vessel with their little sails. Our captain was in the gangway, looking and we didn't know what minute he would say, "Put your dories out." But he didn't that day. Shortly after we arrived home a bunch of us fellows were in the Salvation Army one Sunday afternoon when all at once we saw people leaving. This was because one of our vessels was just coming in with her flag at half mast. It was the very vessel we were watching on the Grand Banks, the day we did not fish. Two men from it swamped their dory and were drowned[13]. They belonged to Jacques Fountaine, Fortune Bay, and one of them planned to be married shortly.

There is only one dory skipper

Even a dory skipper in this deep sea fishing business had quite a responsibility. There are two men in a dory and one of them is called a dory skipper, and he is supposed to see that the dory is well fitted with gear, drinking water, bread sealed in tins, etc., and he is supposed to be experienced. Many times he and his dorymate are in grave danger out there on the Atlantic in a little dory. It's a "must" for him to know the compass. Some men have gone to the banks for years and never could learn the compass, but these are very few. Sometimes both men in the dory are experienced and one is just as good as the other, but still there is only one dory skipper.

Before you left the mothership to set the gear in the water, you threw overboard a rope fifteen or twenty fathoms long with the end made fast to the vessel, and whichever way the current ran the rope pointed in that direction. If there were eight or ten dories, almost half had to go in the direction of the current, away from the vessel and set their gear towards the ship.

Sometimes it happened that you had four tubs of gear to set and you went away and set towards the vessel, but did not go far enough, and when the gear was set you had half a tub or even a whole tub left. You had no choice, only to bring it on board the vessel, and, believe me, the skipper was there to greet you with a few choice cuss words.

All you had to run for was a small buoy

When you left the vessel to haul the gear, one-third of the dories went to leeward the length of all the gear, sometimes two miles, but at least a mile and a half. There is thick fog on the Grand Banks ninety per cent of the time, plenty of wind and a big sea running, and all you had to run for was a small bouy on the end of the gear, with a little marker called a black ball. Lives depended on that; if you missed that little ball, there was only one thing to do: row back to the vessel. And many have been lost. Some men have been astray for days and have rowed 100 or even 200 miles before reaching land. Many lives have also been lost by overloading the dories, especially when counting fish.[14]

One day we were on a lot of fish and Captain Smith kept

three dory crews — six men — aboard, dressing down the fish, while the other eight went on out, fishing.[15] My uncle and dorymate was a splitter[16] and he stayed aboard while I went in dory with a big man from Garnish. Ches' White was over 6 feet tall and 230 pounds, and a wonderful fisherman. I don't think he was ever beat counting fish in his life. What a man!

We went out that day and underrun our gear, and we loaded our dory. And he used to load his dory some deep! There was no big lot of wind, but it was loppy and we had so much fish that the head of the dory was going under as we hauled towards the schooner. He was in the bow rowing, and he said, "Archie, you come in the head of her and I'll get back aft because I'm heavier than you and that'll rise up the head." I don't know how we got back aboard that day, but it took a lot to sink a dory.

After we discharged our caplin trip we got ready for a squid baiting. There was always a bit of enjoyment getting squid, especially if the squid was scarce. Then it would take two or three days, and, if you were lucky, you might strike garden parties and dances ashore, and that was quite a pleasant break for us. We finished the voyage around the first week in October. The price for fish was very good during the past two years, and whoever was lucky enough to get a good voyage did well. My brother and another member of our crew didn't finish the voyage. They were taken with typhoid fever and had to be rushed to the hospital in St. John's. The crew was landed at their homes by the middle of October.

In December, I spent a night with a friend at Stone's Cove, where the banking vessel *Warren M. Culp* was moored at the wharf. That night, no one was on board when she broke adrift and drove on the rocks close to the house where we were sleeping. Men were called out of their beds before daylight to get in their dories and pull that large vessel off the rocks. She was badly damaged before skill, perserverance and a lot of hard work saved her. She was taken to St. Pierre to be docked. But it was a frosty winter, and the slipways there were frozen in, so she did not return until just before the spring fishery began. This vessel, purchased from Lunenburg a couple of years before, was the first deep sea fishing vessel (perhaps at Stone's Cove) to have a deck engine for hoisting sails and heaving in the large cable and anchor.[17]

But there were always tragedies. Many lives were lost on the bowsprit. Because the jib is the first sail you have to shorten when the wind increases. If you drive it out too long, it was a dangerous place to put men out on. Many times bowsprit and men went right in under. One Grand Bank skipper lost three or four men on the same trip that way. Four men were lost from Grand Bank vessels that year. In May month two men from the *James and Stanley* were drowned, and another two went astray from the *Preceptor* and were never heard from. Another Grand Bank vessel, the *Bella Bina P. Domingos*, was lost on the Labrador coast in the fall of the year.

During the year, I became very interested in the large three-mast vessels sailing from Grand Bank. Some were commanded by quite young men, so I'd been thinking of going in the foreign trade. In January 1920, I secured a berth on the Grand Bank three-mast vessel *General Horne*, and her owners told me to be ready to join her around the first of February. I had to go to Belleoram to take the steamer *Glencoe* that was coming down the coast, but when I reached Belleoram, I found that she was stuck fast in Gulf ice and could not get down. I informed the owners by telegram and the *General Horne* waited two days for me. But it was impossible for her to wait longer, so she sailed without me and I had to return home.

By this time most of the vessels had their fishing crews and I had to find some way of going fishing to the banks again. I spent quite a bit of money on telegrams, and I was beginning to get discouraged. March month was slipping away and all the men had gone. But luck followed me. A Captain Anstey came to Rencountre East, where his vessel, the *Dauntless*, was moored, and he needed a couple of men to go in dory. He heard I wasn't shipping, and it was a pleasant surprise when I got a telegram from him offering me $550 wages and shares. He came for me the following morning.

I had no idea who my dorymate would be. I found out it was a young man from Sagona. He had been fishing a year longer than I, but he wouldn't be dory skipper. He was perfectly satisfied to go with me, and said, "We'll make out alright." When the voyage was over we both had our wages made and a little more. We enjoyed the year's work and got along well together. It was a real experience for both of us.

There was a big slump in the fish market at Portugal

There was a big slump in the fish market at Portugal in 1920, and the company we were with (Samuel Harris Ltd.) had a number of their three-mast vessels at Portugal loaded with fish, with no possibility of getting rid of it. It was a real crisis, especially for this company, because they had a large number of foreign-going and deep sea fishing vessels. They claim the price of fish in Portugal went right down because the market was flooded with cheap Labrador fish. Harris and Company couldn't even sell them. They had to give it away. So when settle up time came it was impossible to get all of our wages because of the money shortage. We hoped to get the remainder as soon as the money situation improved. When the voyage ended the fishermen were taken to their homes.

There is always a mixture of joy and sadness in this fishing business. The vessel that I was supposed to join in January, the *General Horne*, commanded by Captain Rogers, was returning from Portugal en route to the northeast coast of Newfoundland. She never reached her destination. She was lost with all hands. Six months after she was officially given up as lost, one of the Grand Banks vessels sighted her driving around the ocean bottom up.

In 1921 I decided to make a change. Instead of going to Grand Bank, I joined the *Flora S. Nickerson*, commanded by Captain Harry Pope from Stone's Cove, and my older brother and I were in dory together. We went winter fishing in January on the southwest coast in the Rose Blanche area. It was a severe, frosty winter, and shortly after we arrived at Harbour Le Cou, all the vessels were frozen in, jammed with Gulf ice. We got on the fishing grounds only for a few days that winter. We arrived home the last of March with the hold of the vessel almost empty, and we never made enough money to pay expenses. But we mustn't give up hope.

After a few days, a herring baiting was secured and we left for the Grand Banks. Our voyage ended in the first week of October with a small catch of fish and we made only $300 for the year's work.[18] A few days after we arrived home my aunt passed away at the age of forty-eight, leaving a family of nine children.

Twice . . . we worked seventy-two hours without sleep

I took my younger brother in dory with me in 1922, another year of responsibility for me. We joined Captain Bill Evans in the *Carrie and Nellie*, which operated from Epworth, Burin. William Forsey, a Grand Bank merchant, owned the vessel and had a premises there. We worked very hard and counted fish all of the time. Twice in particular during our trips we worked seventy-two hours without sleep. Fish was plentiful and the captain's idea was to bring on board all the fish we could by day and dress it by night, which meant there was some fish left on deck when daylight came. Even after the first twenty-four hours everybody was getting tired and sleepy.

It was stupid to drive men like this and it had its disadvantages. Some of the fish was rotten before it went under salt. You were bringing in a lot of fish, piling it on deck and rotting it. Then, next morning, you'd get up and go away, and every morning there'd be more on deck. You couldn't do anything. And you can't get as much work out of tired men as you can from men who have had a fair amount of sleep. Besides that, you're taking crazy chances, because men out in their dories could fall asleep and never be heard of afterwards. If you should happen to close your eyes, that was it!

You'd go dead to the world. Then, out in a little dory like that, perhaps about loaded with fish, a little sea could come over and swamp you just like that!

When we discharged our spring trip at Epworth that year we were saddened by news about the death of one of our young men from Anderson's Cove. This relative of ours was only twenty-one, and mate on a three-mast foreign-going vessel from Grand Bank. He was killed when accidentally caught in a deck hoisting engine. It sure cast a gloom over our little settlement.

We never had a statement of how much was bad fish and how much good

My brother and I made an average voyage this year.[19] But we received small wages because of bad fish which only fetched West Indies prices. It often happened that we got a trip of fish, came in, washed it out and put it on the beach, then went out and fished another baiting, and upon return we found that they hadn't started to dry the fish caught on the first bait. It depended on the sun. So the fish went bad and had to be sent to the worst market — the West Indies. Half the cash from this fish belonged to us, the fishermen, but we never had a statement of how much was bad fish and how much good. The owner just gave it in a lump sum: $100 or $1,000 worth of bad fish. We wouldn't get it in quintals or pounds, so you couldn't know.

During the past two years I had a strong desire to become a captain. I knew I wasn't capable yet, but I was working toward it. I liked to look at the captain's chart when I got a chance, and one experienced man who had been captain before showed me how to take off a course for my first time.

In the first week of December that year I was home one morning and decided to walk to Stone's Cove. When I got there I went in to my friend, Albert Pope's house, and he said, "Arch, you come in and have dinner with us, now, before you goes down today." It was Saturday, and they had pea soup for dinner. We were sitting around the table when someone bolted through the door and said a man had fallen through the ice and drowned in the pond. Another man was walking out and saw it happen, but there was nothing he could do. He ran straight out to Stone's Cove, a couple of miles, and all the men took off like a jet.

I went with them, and we beat the ice off a dory and carted it up over the hill and went in the pond. One man took a cod jigger, and I can see the man now. They hooked the jigger in his nose and he came up. It was a sad affair. He wasn't very old, in his forties, and he left a wife and family. He was in the Orange Society at Stone's Cove, so they carried his body up to the Lodge and kept him there a couple of nights. Then they had a funeral on a clear day. It was a sad affair just before Christmas.

The winter of 1923 was very severe. In early February and March the Arctic ice and the Gulf of St. Lawrence ice met, and the Grand Banks of Newfoundland were covered. The entire southwest coast and Fortune and Placentia Bays were blocked for almost three months, and people could walk from Belleoram to the bottom of Fortune Bay over ice. All shipping was at a standstill and food was scarce in many communities. Men caught herring through holes in the ice for food, and it was often said that herring never tasted as good before or since.

Vessels coming from the European countries had no warning until they were solid in the ice on the Grand Banks, and many Grand Bank vessels coming from Oporto ended up at Halifax and had to wait there until the ice broke up and moved away. Fishing vessels were also delayed going on their spring trips, and vessels that went to fish in the Rose Blanche area were jammed in ice all winter.

We ended up high dory, counting fish on our first trip

This year my uncle, Wilson Green, and I were dorymates in the *Dorothy Melita*, a Grand Bank vessel owned by John B. Patten.[20] The captain was my first cousin, Reub' Thornhill. The ice moved out of the bays around the first of May, and the

S.S. *Glencoe* came and collected all the fishermen. Even though the vessels were late getting away on the spring trips, most of them ended up with big trips. It seemed as if fish were more plentiful than ever on the Grand Banks and Western Banks. Some scientists said it was because there was so much drift ice all over the banks. We had a crew of fine fishermen on board, and I guessed there would be a few extra lines used. My uncle was dory skipper that year and very smart baiting gear. We ended up high dory counting fish on our first trip. Only two dories are most talked about when fishing: one is "high," the other "low."

After the caplin trip was discharged the vessels went around looking for squid for bait in such places as Pushthrough, Rencontre West, and the Burgeo area. Other vessels went in the opposite direction, some to Placentia Bay. That's where Captain Reub' decided to go, but squid was very scarce in that area, so scarce that some of the vessels that struck the squid were in with a baiting before we got on the fishing grounds. This didn't go over very well with the owners even though our skipper was an eager fisherman. He received a telegram from them. What it said, we will never know, but he told us to put the sails on her, and although it was a bad night, he walked the deck and never let up until we reached Grand Bank. He was there a few hours and then we left for the southwest coast, where we baited and sailed for the banks. But the weather was bad and fish scarce, and we secured only three hundred quintals. It was early October by the time we discharged our catch and it was rumoured that we wouldn't be going out any more. But we did and we came in October 20th with another eight hundred quintals, which gave us a very good voyage.

I began to feel as if I was getting somewhere

Before the voyage ended, the captain asked me to go second hand (mate) with him next year. I was very pleased and accepted, and I began to feel as if I was getting somewhere. But I also knew it was going to be extra work. A second hand on a banking vessel is supposed to keep the vessel in good condition, keep all ropes and wires repaired, and any work aloft is supposed to be done by him, whether it is stormy or calm. He is also responsible for the vessel when the captain is not on board, and supposed to be in the bow of the dory, boss of hauling caplin with the seine, and many other extras as well.

In January 1924 we joined the *Dorothy Melita* again to go winter fishing on the Rose Blanche. My cousin, Frank, and I went as dorymates. We had some of the best fishermen on board, but we had made up our mind to try and keep up with them. To be highliners you had to work and use all kinds of stunts. If the captain gave orders to use 40 lines of gear, and I'm smart enough to get up to the cabin in the morning and can be finished baiting my gear half an hour ahead of the rest, I'm going to put on two or three extra lines. But no one, especially the skipper, is supposed to know that.

When our vessel is anchored and the gear left out overnight, and we're underrunning it, if I'm on watch at night, I'd sneak down in the cabin and get a couple of extra lines of gear. You'd haul open your dory sail, and roll the lines up in it. Then, when you go out, you put up your sail to sail down, and, while the dory skipper is steering along, his dorymate is baiting up the extra lines. You'd put them on the outside end of your gear and catch more than double what you'd get near the schooner. I've seen a fish on every hook on the outside end! But if you put on extra lines and were behind coming back aboard, or behind baiting up your gear, you're not able to handle your gear. If you can use it and be on time with the rest of them, it doesn't matter to anyone.

I remember another stunt my cousin, Frank, and I used on the *Dorothy Melita*, and a good many other good fishermen used too. One day, we were fishing out on Mizzen Bank and caught 800 fish for our dory. Five or six of the other dories had scattered fish on only half of their gear, while the other half of it went out over the shelf into deep water and caught only black dogfish. That evening, when all the dories were on board, the skipper hove up and anchored again on the outside edge of where Frank and I were that day. Now, when you weigh anchor, even if it only comes off the bottom, you have to draw new courses for your dories and can't go on the course you had just before. We drew a course further up and knew that some dories will still go into deep water, but we didn't say anything to anyone.

If we went in the direction of our new course, we also knew we'd get no more than half of what we had yesterday while the dories fishing on our old course would come in with 800 or 1,000 fish. When we got in our dories again, everyone else went on their courses. We gave them a chance to get a bit ahead of us and then moved in between two dories, going right back on our old course.

You had to work, by God!

The skipper could see us, because there was no fog. And when Frank and I came aboard, he said, "Arch and Frank, what did you fellows do today? Your course was up there, wasn't it?" "Oh, yes, sir." "Well, why didn't you go up on your course? You was down there between two dories, takin' up more space. You fellows had no business to do that!" The other men, when they came on board, were wild about it too, especially the two dories we set between. We were taking a lot of fish they were going to get. "Well, sir," we said, "perhaps you'll see this evening, when we get our gear in, that there is nothing to get mad about."

After we had our snack, we took in our gear and brought aboard exactly the same amount of fish we had the day before. Sixteen hundred fish for two days! And, work! You had to work, by God. If you didn't work, you wouldn't get any fish. That's all. The other fellows got fish, but we happened to have the most. Six or seven dories had 780-790, but we had 800 and were top dory. About half the dories had a good day's work for the vessel. But that's it, counting fish. There were some dories that had less than 200 fish that day.

When we came aboard that night, they had to do the same work, putting down all that fish, and that's how it went on.

We had frost and gales most of the winter, and when we arrived at Grand Bank our trip fetched only 400 quintals. After discharging our winter trip we baited and sailed again for Bank Quero. We fished our last two baitings on the Grand Banks.

. . . narrow escapes are common

The Atlantic is very dangerous at times in a small dory, especially when the sea is rough, and narrow escapes are common. One morning in particular we all left our vessel to take in the gear and we had to haul to windward. Before we had hauled half of the 26 lines it was pretty rough. After we hauled our gear we put the little sail on the dory, but it was too much. We had to tie down one third of it and run before the seas towards the vessel. Just before reaching her, a big sea hit us and almost filled the dory. The two men in the dory behind us saw what happened and thought we were gone. They quickly hauled the sail off their dory and rowed toward us. We kept cool and got the water out of the dory. We were fortunate to save ourselves, but not too far from us they weren't so fortunate. Just as we were all safe on board and homeward bound we saw the *Bessie MacDonald*'s flag at half mast. We went alongside and the captain said two of his men swamped their dory. One man was saved by clinging to the dory bottom. The other man, a Mr. Miles from Little Bay East, drowned before help could reach them.

After you reach Grand Bank and the work is finished for the day you usually "dress up" and go for a stroll. We always look forward to a Sunday in port. The Grand Banks is a dangerous place for a sailing vessel because there are all kinds of ships coming and going, fishing vessels, steam trawlers, large passenger ships and freighters, etc., and the fishing vessel had only a small fog horn to signal with. Many vessels and dories were run down in foggy weather and many lives were lost.

Our voyage ended without too many incidents. One man got typhoid fever and had to be taken to North Sydney. On the way we struck one of those August gales and were delayed twenty-four hours getting him to a doctor. Some vessels were lost on Sable Island, and some lives were lost in Placentia Bay. My dorymate and I did well counting fish and ended up with a good year's work.

That fall the owners decided to send the vessel to Prince Edward Island for a load of produce. But I managed to get home for a few hours to see my family. My sister had been seriously ill with typhoid fever, but she was getting better. Six of us who were fishing would be going on the vessel to Prince Edward Island. It was the first time there for most of us. I was classed as mate now and we were freighting instead of fishing. We were all on wages; the crew got $35 per month and the mate $45.

We returned to Grand Bank and discharged some of the cargo there, and some at Lawn. It was December when I arrived home with plenty of vegetables and other things for the family. The captain asked me to go fishing with him again next year as second hand, and I would have the same dorymate.

Father and son were never seen again

The next season, 1925, brought sadness to many families when the vessel *Rex* was lost one Sunday morning, June 28th. She was lying at anchor on the banks in a dense fog. The crew was on deck, baiting their trawls and waiting for the fog to clear. The cook's ten year old son was fast asleep in his father's bunk in the forecastle. Suddenly, out of the fog came this 17,000 ton Cunard liner. She sliced into the *Rex* on the port bow and cut through the vessel, her bow coming out of the starboard side just forward of the main rigging. The foremast was ripped out and ground into small pieces of wood. Forty tons of ice that was in her hold for icing fish came floating out and the men who were on the starboard side were thrown in the water among the ice and had no chance to save themselves. The heavy cable holding the anchor was churned into small strands when the liner cut her way across the deck. The cook, who was on deck at the time, jumped down to the forecastle to save his young son, but father and son were never seen again. The liner put out lifeboats and picked up nine of the crew and the captain's body. He may have died of a heart attack in the crash.[21]

But 1925 was another successful year for us at the bank fishery, and we now had three good voyages in succession. That fall the skipper and I, as mate, were to sail the schooner to Prince Edward Island for a load of produce after we gave up fishing. But shortly after we arrived at Grand Bank to discharge our last trip of fish it was rumoured that another company there, Forward and Tibbo Ltd., was trying to persuade our skipper to sail with them. This often happens if a man is a prosperous fisherman, whether he is skipper or in dory. And in this case they were offering an almost new, more modern vessel, the *R.L. Conrad*, a large knockabout. They wanted him to got to Lunenburg and bring down the new schooner, and he would take her fishing the next season. About three days before we were to sail, Captain Reub' decided to go with them.

I was determined to get out of the bow of a dory . . . .

I had been mate with this skipper now for two years, sailed with him for three, and my dorymate and I did well. So I wondered if the skipper would recommend me to take charge of the *Dorothy Melita*, but no such luck. Perhaps it was for family reasons that he didn't recommend me. But I said to myself, "Here's a chance for me to be skipper. I've already had a hard struggle trying to get to the bank fishery, so I might as well be prepared for more of the same." I was determined to get out of the bow of a dory, so I decided to recommend myself. I was now twenty-four years of age.

I didn't say anything, didn't ask the skipper a thing, but waited until he left the vessel and went on to the other company. We were sailing for J.B. Patten and Sons Ltd., at the time, so I went up to our manager's office — Mr. J.B. Patten, "John Ben," we used to call him. I said, "I understand Skipper Reub' is not going to the island now." "No, he's not. He just gave in his notice." "Well, I was wondering, sir, if you'd consider giving me the schooner to go up for the load of produce?" He looked at me — he didn't have too far to look, because I wasn't very tall, and he said, "Well, Arch, I think you're a little too young yet. But I'll give you credit, having the guts to come and ask for the vessel." I told him, "I am as old as Skipper Reub' when he went master." And he said, "I think you'd better keep on for another while yet. You got plenty of time, you're only a young man.

We got a retired captain here now, and he's more experienced, so I think we'll give her to him." Mr. Patten had my record in his office and knew all about my work. When he asked Skipper Reub' about me, he said I was a bit young and inexperienced, and he recommended his brother, but he didn't get the vessel. I asked a second time, but Captain Joy Hyde, who was used to those trips, got the vessel and I went on as mate. It was another setback for me. We went up to Prince Edward Island and came back down, discharged and tied up the *Dorothy Melita* for the winter in Grand Bank. Then I went on to my home in Anderson's Cove.

In early March 1926 we left home to join the *Dorothy Melita* at Grand Bank. I was still mate. I didn't believe in running around, and thought I'd stand a better chance to go skipper by hanging on to the same company. Our new skipper was a stranger to me. He had been foreign-going in three-mast vessels most of the time and hadn't done much deep sea fishing. Unfortunately, he was a kind of laddio — a heavy drinker, and things didn't go so well.

We left Grand Bank on March 17 and touched in at St. Pierre to buy a few things. We sailed again the next morning and we were half way out on our fishing ground, the Western Banks, on our first bait when we discovered our fresh water tanks had burst out, leaking, and we lost our drinking water. Sometimes we'd be out three weeks or a month and we had to have a good supply of water. So we went back to St. Pierre to get the tanks repaired and filled.

He went ashore on a real bender

After we anchored in St. Pierre harbour, we didn't see much of our captain. He went ashore on a real bender, and it was ten or twelve hours before we saw him coming. Before he reached the vessel I saw him going with his hands, meaning to say, "Hoist the sails. Hoist the sails. Be all ready." But when he got in close range he shouted, "Hoist them Goddarn sails!"

We were eleven dories, twenty-two men and the cook and kedgie, and when we heard that, the men put three dories out — enough to carry every man, and rowed ashore to St. Pierre. The men decided that a man who drank as much as he did couldn't be responsible for their lives on the fishing grounds, so we left her right there and then. We weren't going to sail with him anymore.

I felt sorry for him, because he was in a poor fix then, and he was a good man otherwise and a number one seaman. A little after we landed at St. Pierre, he came ashore to where we were, crying to us, telling us he was sorry. He promised he wouldn't be drinking anymore, that he'd be a different man if we would give him another chance. I told the men I thought it was alright to go back, so we went back aboard, had our tanks repaired, and this time we sailed for the Grand Banks.

Early March is a poor time to go to the Grand Banks because the weather is often bad. But we went there anyway, struck bad weather and had to shorten sail the first night out. At twelve o'clock that night she was hustling, a big storm was coming on, and we thought he was drinking again. He broke his promise in less than twenty-four hours. I went down and called him, but I couldn't get any sense out of him at all. So I gave the crew orders to take in the rest of the sail. I never went below that night. I was really in charge of the vessel.

Come daylight he came up, looked up — he was a little sensible then, and said, "Take in the jib. Take in the mains'l." I told him everything was in and snugged away, and that we were hove to on the fores'l. "Who in hell gave you orders?", he said. But he went down in the cabin again.

The trip lasted about a month and we never got a dory fishing, because nearly every man aboard got the flu.[22] I was sick for thirteen days and never got on deck, and he was drunk almost every day. Water and food began to get short again, so we left for home. When we arrived at Grand Bank with no fish, he told us not to leave the vessel, because he was leaving. All the crew were going to leave him. We couldn't get along, we weren't used to men like that in fishing.

I thought the owners might give me a chance to skipper the vessel now, and every man indicated he would sail with me. I looked like a ghost from being sick so long, but I went up to Mr. Patten in his office and said, "What about giving her to me? I can do it. If you'd only give her to me, I'll go out and get some fish. The men will all go with me." But he still thought I was too young. "There's a man up in Fortune Bay, a skipper man, he's been skipper for years and years. He's home this spring and isn't fishing. We can get this man to go aboard. I want you to take the vessel and go up to Stone's Cove and get him." Perhaps it was the best thing that ever happened to me. We sailed with this man, Captain Adams, for the rest of the voyage and he was a good man. The voyage ended with a fair trip, but not nearly as successful as the last three years in the same vessel.

We went around the bay and landed all our men from the *Dorothy Melita* with their provisions and, so we thought, snugged away for the winter. But my cousin, John P. Thornhill,[23] in Grand Bank, decided he would go to North Sydney to pick up the *Freestead*, a small, four dory schooner of about forty to fifty tons, and take her fresh fishing. She carried about ten men; eight in dory, and the skipper and cook. He wired and told me what he was doing, and asked if I'd be interested. I wired back that I would, and got my dorymate, and we left home in November.

This was new to me, as I had always been salt fishing. We fished the last two weeks of November and on until the first week of January. The weather was very cold and it's much different from salt fishing. We'd go out, catch our fish, and land it fresh almost every second day, and were paid almost as soon as it was discharged.

We spent Christmas Day at North Sydney, and were on our way to the fishing ground again before daylight the next day. We landed a good trip of fish as the year 1927 came in, and finished about the middle of January. The ice began to come up out of the Gulf then, and you had to give up fishing. So we tied up the schooner. We did extra well, made a nice bit of money for being gone only two months.

I told him how I had tried and failed to get the Dorothy Melita

At that time we had to take the old *Portia*, a ferry, across the Gulf from Sydney. But we had to wait a week in Port aux Basques before the steamer made connections to come down the coast. While we waited in Port aux Basques, my cousin, Captain John T. Thornhill, said to me, "Arch, you know, you're capable of taking a schooner now. You're twenty-five, almost twenty-six, and I've recommended a good many young men in my day. They've all done pretty good, so I think you're capable for a schooner." I told him how I had tried and failed to get the *Dorothy Melita*, and he said, "That might do you good." Skipper John had the *Vera P. Thornhill*. He had her built down in Shelburne, Nova Scotia. Then he told me he wasn't going fishing in her next year. "I'm going fishing in the *Freestead*, so the *Vera P. Thornhill* will be wanting a skipper next year." He didn't say any more.

I arrived home at Anderson's Cove on January 20.

Chapter Four

My First Command

Arch's story brings us to a major threshold and turning point in the careers of many banks fishers. Now age twenty-six, he is an established mate or second hand, and the dream of becoming a fishing captain burns bright. But what does it take to be chosen? The answer lies in the convergence of experience, demonstrated ability and eagerness to command, opportunity, and support from key influentials or gatekeepers — vessel owners, and especially fishing skippers like his cousin, the big fish killer, Captain John T. Thornhill. Selection occurs within a network of kin and patrons. Once chosen the captain is expected to overcome the sea's indifference and all operational uncertainties, such as crew recruitment and management, bait and fish scarcity, crew illnesses and accidents, miscalculations, equipment failure, and bad weather and ice. His vessel's voyage must bring profit.

A central issue for every captain, of course, is: How do you handle a crowd of men? Individually, men recruited to fish varied in age, experience, competence, eagerness, and familiarity with the skipper. How does a fishing master maximize their productive potential without undue risk to authority, his men and vessel? In this chapter we learn how fishermen, Nova Scotians, on the one hand, and Fortune Bay Newfoundlanders, on the other, employed the same banks fishing technology on the same grounds, yet organized their production — their work or labour effort — very differently.

In this, his first command, a Nova Scotia-owned banks schooner he crewed mainly with Fortune Bay men, Skipper Arch decides they will fish on the "count" earning system that had steeled him. But they oppose him and prefer the less onerous, more transparent (average) share system used by Nova Scotians. The conflict challenges a young captain's authority.

No matter why a man "left the turf" and his place in the climb to captain, banks fishing was only one part of his life. Family, friends, and activities ashore were compelling interests. Men married, established their own homes, and had children. But their long absences away to fish the banks imposed the major burden for managing homes, childrearing, and family finances upon wives ashore. The ideal wife was an industrious contributor and manager of her household's resources. She was frugal and skillful at "making do". Women always had to brace for hardship.

It was not easy to see their men depart for the banks. Everyone knew that some never returned. And widowhood usually meant the penury of the dole, charity, and backbreaking work on a merchant's flakes or beaches. Arch's wife, Ruth, was an experienced, college trained school teacher. But, should worse come to worse, a teacher's meagre monthly salary was not sufficient to support a fatherless family.

Most wives of banks fishers came from outport family homes. They had learned about hard work, and the skills and values of self-sufficiency and self reliance. They were expected to manage their fears and lonely responsibilities. Many were "nervous" for good reason. It took special strength to be otherwise.

Arch recalls how his vessel's Nova Scotian owners chose to withdraw her from fishing after two voyages under his command. He became a fishing skipper without a vessel, newly married, and bound to a home mortgage in a new community. We may guess at his personal anguish. What was he to do? Ruth gave him confidence as he returned to banks dory fishing, perhaps with new resolve. No matter how hard they worked, banks fishermen never knew if a voyage would be successful. Falling fish prices during the 1930s made voyage outcomes more uncertain and inadequate. After settling up many men went home in debt to merchants. Fishing firms failed as their operating costs exceeded market returns, freights declined, and cash jobs grew more scarce. The formerly diverse, dynamic, and optimistic fisheries atmosphere of Fortune Bay earlier in this century was over. The grim conditions and enduring sense of exploitation by fish merchants helped rationalize the old and widespread practice of petty smuggling from Prince Edward Island and other foreign ports where goods were much cheaper than at home. It seemed only just and necessary to make life easier.

Little wonder that men with dependent families especially, looked beyond banks fishing and other customary production, like wood cutting, hunting, and trapping, they managed between trips and voyages. It is an old story of generations of Newfoundlanders driven to seek their fortunes far afield. Arch was unsatisfied to "lay around" between voyages. He went freighting for a more dependable, monthly income. And soon he considered abandoning banks fishing altogether.

1927-1932

When I arrived home from North Sydney, I went to Little Bay East for a vacation. The new teacher I met on my arrival home from the banks my first year at sea was now teaching at Little Bay East again. She had taught school at my home for four years, and one year at Little Bay East.[1] Then she went to university for three years, at Berea, Kentucky, in the southern United States. After graduating at the end of May she returned to her home in Pool's Cove, before going to Little Bay East. We'd corresponded ever since she left, and it was four years since we'd seen each other.

I guess he sized me up

While in the boarding house there I received this telegram from Captain John Thornhill, at Grand Bank, asking me if I would buy his two shares in the *Vera P. Thornhill*, and go as her captain, deep sea fishing. She was a practically new, 170-ton, eleven-dory vessel, owned mainly by shareholders at La Have, just the other side of Lunenburg, Nova Scotia, and was moored up for the winter at Rencontre East, Fortune Bay. It was quite a pleasant surprise. I never expected such a telegram. The previous winter was the first time I had fished with Captain Thornhill, and I guess he sized me up. I was $100 short of the price of the shares, but I wasn't very long borrowing that amount from a very close friend. And I accepted the offer.

As usual, the news wasn't long getting around. I went to Little Bay East in my father's motor dory. After a few days I left with a cousin who had shipped with me before. We took the open dory on a cold February day and went right around Fortune Bay to select a crew of twenty-four men, including cook and dress crew. We needed the dress crew because the owners in Nova Scotia didn't want you to count fish. I'll say

Schooners in port for bait, at anchor, drying their sails, at Burin, c. 1929

more about that later. The weather wasn't pleasant by any means, but we ended up back at Anderson's Cove a few days later with a full crew, many my former shipmates.

I was anxious to see my vessel, my first command. Within a few hours I left home in the dory again bound for Rencontre. I had seen the vessel many times in the past few years, but it meant little to me then. On the twentieth of February I joined the coastal boat *Glencoe* and went to Grand Bank to settle legal matters about my shares in the vessel. The magistrate there arranged the bill of sale, as if you bought the schooner.

In the first week of March I took the vessel from her winter mooring and went around the bay to pick up my crew. I was twenty-six years old on February 27. We went from there to Grand Bank to fit out for the season's fishing. Because the *Vera P. Thornhill* was owned at La Have, Nova Scotia, I had to get supplies from a company in Fortune Bay. I chose Grand Bank as my headquarters, contacted the owners at La Have, and they made arrangements through the banks for disbursements, etc.

You'd say to yourself, "We won't have a storm"

We sailed from Grand Bank with full supplies for my first time in command, bound for the banks, on the 17th day of March, 1927. We touched in at St. Pierre, of course, just long enough for the men to get a few things. There are plenty of storms in March and April, and we went up to the Gulf, which is a poor place to be at that time of year. But you'd risk a lot of things because it was a good place to fish. And you'd say to yourself, "We won't have a storm." I knew I'd taken many a big day's worth of fish on the banks west of Sable Island when I was with my cousins and the other skippers. And I knew, when I became skipper, if fish were there, I could get them.

The weather was good for the first few days' fishing, and we had one very good day with 100 quintals. Then, along about one o'clock in the morning, a big storm came on, and I had to put her head to the north. We got into the ice, this heavy Gulf ice, and a gale of wind. The sea's strife where we were is always bigger coming from the south, and I wanted to put her around, to head the swell heaving in towards us. We could have been gobbling like a duck then, but for the ice. We had only the fores'l up, and there was nothing else to do but try to jibe her over. I took the wheel myself, watching my chance in the ice, the best you could in the night. And when I jibed her over, darned if we didn't break our fores'l boom right off.

We were darn near lost with the rest of them

"Well," I said to the boys, "that's not the first one to be broken, and it's not the first one that's going to be fixed up." So we dug in. We didn't mind that, because we're used to breaking booms and one thing or another. We patched her up and got the fores'l up, and I had the job to try to get out of the Gulf again. The storm went on about two days and two nights, and we jogged around, in and out of the ice. A schooner like that can jog fast, and soon I saw that we were heading up towards Sable Island, where hundreds of schooners have been lost with all hands.

I was getting worried, but I didn't let anybody know until I couldn't keep it to myself any longer. But the crew knew before I told them. She was ranging ahead so fast, and I couldn't get out of it, the way the wind was, and you couldn't carry any sails with the strife of wind, blowing so hard. That night she went through water on Sable Island bar so shallow at one point that the sand washed on our deck. She didn't strike, but we were darned near lost with the rest of them, and happy to come out of it safe.

After the weather improved, we picked up our position from another vessel and started fishing again. Our first trip was successful, about 500 or 600 quintals, which was above average, and I was very pleased. When our bait was used up we returned to Fortune Bay, took on another supply, and returned to the Western Banks.

On our second trip, the weather improved all the time, fishing was fair, and we were heading for a good spring trip. From our first frozen baiting we had 500 or 600 quintals, and by now, on our second frozen baiting, we had almost 1,100 on board. There would be two more baitings used on this trip, if all went well.

Then things went badly. I had sent all the men out to take in their gear on the second day, because the place we were anchored was getting fished out and we were going to move to a new berth, a new ground. Three or four dories were still out. The fog was thick and it was getting late in the evening, when I noticed the vessel had fallen broadside to the current. At first I thought the current had changed, but our cable had been cut off on the bottom and we were drifting away, drifting from the men. All the bank fishing vessels carried both a cannon and a fog horn for use when the vessel is anchored and the men are out on their trawls. We put the cannon up on the bow and steadied it with anchor chains.

I was getting . . . rattled, afraid I would lose two men. . . .

While one man was working the fog horn, I kept filling the cannon with powder and blasting at short intervals. You could hear that sometimes five or six miles away. The dories kept coming after taking in their gear. Finally, one dory never came. I was getting a bit rattled, afraid I would lose two men after such a short time in my first command. I loaded the cannon again, put the match to the fuse, and from that time after I don't know what happened. I was lighting the fuse when I was thrown from the bow to amidships. The first thing the men saw through the fog was my cap blown off from the side of the vessel. They thought it was my head.

A few seconds more, and I wouldn't have had to fire the cannon at all. It was nearly dark and the two men were close enough that they saw the flash of the gun through the fog. After they got on board, they said they heard the sound of the fog horn a few minutes before and took their bearing.

I was sure in bad shape. Blood coming from everywhere and totally blind. I'll never forget the thoughts that went through my mind during the next eight days. They got me down to my bunk in the cabin, and I gave them orders to get me to a doctor as quick as possible. My second hand was a good seaman, but he couldn't even take off a course. He knew nothing about navigation. So the cook, who had experience taking a vessel around, he took over the charts. We were on a bank at the time, approximately 170 miles from North Sydney, so I told him to take me there. I was in my cabin with a dark screen pulled over, unable to see a thing, and in a lot of pain and there was a bad odor of powder. For three days and three nights I was blind.

We got into heavy ice within fifty or sixty miles of Sydney, and couldn't get in, so I advised them to get me to Louisbourg. But it was impossible to get there. Our only other choice was to go on for Halifax. I could have told them to carry me to Newfoundland, but a lot of things went through my mind after coming to my sense. I figured I'd be laid up for a while and knew it was going to be a hospital case.

If we went to the nearest doctor in Nova Scotia, the vessel could go on for our home port, La Have, discharge our fish, and the owners could decide what to do next.

We got as far as Halifax a few hours before daylight. We were so close, I told them to carry on for La Have, another forty or fifty miles. It was a hard struggle to get to a doctor. When we arrived near daylight, they took on a pilot and went up in the river. I was immediately carried to Bridgewater Hospital. After being there a few days, my sight began to improve, and by the time the fish was discharged and supplies took on board I could see well enough to take off a course on a chart, able to carry on again. But my face was disfigured with powder.

It was really a big setback. Until then, our chances were excellent for a big spring trip. We had nearly twelve hundred quintals for the two baitings and two days fishing. By the time I was discharged from hospital and on board the vessel again, it was time to get ready for the caplin trip.

Other captains had accidents like mine.[2] Soon after my accident, the same thing happened to another captain from Fortune Bay. We were the same age and good friends. But he was less fortunate. He was left blind for the rest of his life.[3]

Nova Scotia vessels had a different fishing routine

We had an excellent catch of sixteen hundred quintals on board after fishing our three baitings of caplin, and seemed bound for a big summer trip. We struck a lot of fish. Talking about summer trips, Nova Scotia vessels had a different fishing routine than our Newfoundland vessels. When they began fishing in the spring, they fished mostly on the Western Banks, around Sable Island, and other banks west of Sable Island. They always went to their home ports and landed their first trip, called the "frozen baiting" trip, then they continued on their next trip, the "spring trip," moving farther to the eastern Banks, such as Banquereau, Mizzane Bank and in the Gulf of St. Lawrence.

When their spring trip was finished they went to their

home port again, discharged their fish and took on supplies for a long summer trip to Newfoundland for caplin and fish on the Grand Banks. When the caplin baitings were over, they went to places such as Holyrood and Burin, wherever squid were reported. Sometimes a baiting of squid was secured quickly. At other times one had to go from Holyrood all the way to Placentia Bay. Some lucky man might be fortunate enough to fill his vessel and have a full load with one baiting of squid, after his three baitings of caplin, and get home before sometime late in September.

We went to St. John's for supplies after our caplin baitings were over. When we arrived, we learned that some Nova Scotia vessels had already baited and gone on the Grand Banks again. There were good reports of squid in the St. John's area, so I decided to stay there and get a quick baiting. But it didn't work out that way. We left after a few days with no squid at all, and sailed to Holyrood, where they were reported abundant. By the time we arrived there, most vessels had got their baiting and sailed for the banks. A few, like ourselves, were left without bait, and we had to sail all the way to Burin. By the time we got our bait there, vessels were on their way in from the Grand Banks with bumper trips.[4]

Our bait had gone bad . . . all of it had to be dumped

We sailed for the banks and, within two days, began fishing. One of our men hadn't felt well for the past couple of days since leaving Burin, and he asked me to take him to land again. I got the man to a doctor, who pronounced it was typhoid fever, and he had to be discharged from the vessel. Our bait had gone bad by this time, all of it had to be dumped, and we went in search of another baiting. The only squid that was reported by then was on the southwest coast.

The summer was flying by and some Nova Scotia vessels were on their way home with their trips. It was getting into August now. We secured another squid baiting on the southwest coast, and I decided to go on the Western Banks this time, as they were much closer than the Grand Banks.

We were returning to land after fishing our baiting, and getting very close to Newfoundland when one of the severe August hurricanes suddenly came upon us. We had no radio warnings in these times. We hove to in a dangerous position. It was one of the worst storms in the history of the North Atlantic fisheries. Five vessels were lost with all hands; one from Gloucester, U.S.A., and four from Nova Scotia, taking a total of 107 lives. The date was August 24, 1927.

The storm had nearly abated, but a big sea was still running. It was night and our position was close off the Lamaline shore of south coast Newfoundland. We put on more sail to get out around, but the vessel struck a reef and within a few hours the water was up to the cabin floor. The pumps were quickly manned, and we managed to reach Burin the following day and immediately docked the vessel.

It was a hard year's struggle . . . from beginning to the end. . . .

Considerable time was lost before we could go after more bait. We'd had hard luck since fishing our three caplin baitings. It was around the first of October when we arrived at La Have, with our voyage finished. It was a hard year's struggle almost from the beginning to the end of my first command. But when I looked back and saw all the things that happened this past year, I thought we were lucky after all. Our catch for the year was above average.

After discharging our fish, the crew took supplies for their families on board and most took a day off to go to Halifax, to get some nice things to take home. We arrived back in Fortune Bay after a pleasant trip and landed all our men and their supplies. I took the vessel with a skeleton crew to my home, cleaned and painted her inside and out, and put our dories ashore. My father made the necessary repairs to them during the fall and winter. When the vessel was in shape for the next year's operations, I carried her to a safe harbour just two or three miles from our home and moored her up for the winter.

This year I learned that Nova Scotian fishing methods were very different from Newfoundland ways. Of course, I had heard about this for a long time. But now, the first time fishing out of Lunenburg, I saw it for myself. It would almost take a book in itself to cover everything, but several things should be mentioned.

First, our Newfoundland fishermen count fish, which is hardly fit to talk about. There've been more lives lost through that method alone — loaded their dories so deep, putting every fish they could in 'em. And then she'd swamp. Whole schooners' crews lost, if it was all put together. Because of the count. Many times I've come back to our vessel with the gunnels just barely out of the water. We'd come aboard and learn some men lost their lives. There are things in your mind, that you would never forget.

Nova Scotia fishermen never counted fish. When they started fishing in the beginning of the season, they carried a "dress crew"; one man was called the "splitter" — in many cases the captains split the fish; another the "header" — to take the head off the fish; another man was called the "gutter" — he took the insides out of the fish; and another was the "idler" — he washed the fish and put it in the hold. In the hold, the "salter" salted all the fish. The header, gutter and salter were always young men who were just starting out, having their first experience in deep sea fishing. And no matter how large the vessel, they carried four or five fewer dories than our Newfoundland vessels. Where we used ten or eleven dories, they used only six.

Most of their fishing was carried on at anchor. The vessel anchored when the captain found the right position. They rarely used more than twenty-two or twenty-four lines a dory, while our Newfoundland men used as much as forty lines a dory. The exception was on our caplin trips, when most Newfoundland vessels fished at anchor.

If you didn't get up early . . . you were a dead duck right away. . . .

Nova Scotia fishermen very seldom lit a night torch when they heaved out in the morning to bait their gear. Because it was already getting light when they got up. Our Newfoundland style was to heave out as early as twelve o'clock in the night. If one waited until two o'clock in the morning, he would be called lazy! This may be hard for the generation that reads this to believe, but I knew skippers, and sailed with some, who had the foolish idea of keeping the clock an hour fast! But there was no such thing as, if you went to work an hour early, you would get to your bunk an hour earlier. I was used to this Newfoundland method then, of course. And, that first year as skipper, if you didn't get up early, the way I wanted, you were a dead duck right away.[5] But the men didn't give me any trouble that first year. That came later.

If you were on a lot of fish, the idea in Newfoundland was to bring on board the mothership all you could and pile the deck full. And many times the stars were in the sky before the last dory got on board. Then, after a quick snack, get to work for the rest of the night, dressing the fish that's on deck.

Many, many times, almost half the fish were still on deck when dawn broke. And then it was, get in your dories, go through the same routine, and pile the fish on what was left on deck. By the time we reached the first fish, it was spoiled, just in good shape for what was called the "West Indies price" — practically nothing! We fishermen called it a starvation price!

Nova Scotia fishermen could have a cup of coffee and something light to eat when they got out of their bunks, bait their twenty-two or twenty-four lines, go and set their gear, then come on board and have their good breakfast. And there was more on their tables than a few leftover beans, like I have seen myself many times. It wasn't until the mid-1930s that we ever got a case of milk for our schooners. We had molasses, but a drop of milk, sugar, tinned fruit, fresh meat and such foods were all called "extras" by the owners. And the crew had to pay for all of it. The owners found the other food, and fishing gear, sails, anchor cable, and vessel from their half of the earnings.

Shortly after breakfast, Nova Scotia fishermen would haul their gear, bring on board their fish, put it on the deck, and as quick as the first fish was on deck from the dories, the dress crew would be hard at it dressing and salting. The men from the dories would have a snack if they needed one, or lay in their berths, play cards, and relax. When the time came, the mate would ring the ship's bell, and as quick as a cat could wink her eye, the men were on their gear again.

They would make four "runs" like this each day. On the last run, in the evening, all the crew pitched in to dress the rest of the fish, which was usually only the last lot brought on board, because the dress crew already had the rest of the day's catch put below. They would have to have a very big day's catch to be up any longer than an hour or so after dark, while our poor Newfoundlander often had no sleep at all for two or three days in a row.

Another Nova Scotia method I admired was how, when vessels arrived from their trips, they always settled up from the previous trip landed. As quick as your vessel arrived, a statement of your past trip was brought on board and pinned on the cabin wall for every man to see for himself. It was a real statement too, giving account of every cent spent before the men's share was taken out, such as expenses for bait, fish-making, and the captain's percentage. The men could read the statement and go immediately to the office for their money. This was very different from what we were used to in Newfoundland, up to 1927.

Before you began fishing, you had to start charging things

In Newfoundland then, everything had to be charged. Even if you had landed a trip of fish at Grand Bank, washed it out and put it on the beaches, it made no difference. From the time before you began fishing, you had to start charging things. Everything, from a needle to an anchor, would be charged to your account. And the price was higher than if you paid cash. There'd be two or three cents added even to the price of a pound of rolled oats, if you didn't have the cash to pay for it. And the higher the item's cost, the more was charged on charge prices! It was clear Roguery![6] And then you think you wouldn't smuggle in a sack of potatoes, or sugar, or something?! A small bit. But for some people, there were only two things that weren't smuggled; anything too hot to handle and too heavy to carry!

In March, 1928, I took the vessel from her moorings and went around to different places in the Bay, took my men on board, and went to Grand Bank to take on supplies and sailed for the banks. This year I decided to go on the Western Banks west of Sable Island, where practically all of the La Have and Lunenburg vessels fish their frozen baitings, and then go to La Have and land our trip. Again, things did not work out as I had planned. First, we encountered gales on our way there. This delayed us quite a bit, and a few days after we reached our destination, one of our men took sick. I had to carry him to Halifax, where he died a few days later.

By the time we reached the banks again, many vessels were already on their way home to land their catches. In order to catch up, I decided to continue fishing and not land before the spring trip was finished. But things went badly for me.

They kind of got the upper hand on me

The same crowd of men were with me this year. Very few of them ever fished out of Newfoundland, however. They always went out of Lunenburg. Although they were Newfoundlanders, a lot of the crew weren't used to the Newfoundland fishing method. And I was still not used to the method used by Lunenburg skippermen. This year the men wouldn't get up before daylight. I was almost the youngest man aboard, and they kind of got the upper hand on me. I did everything a man could do to get them out. But, no, there were too many of them against me. So I decided it would be best to go along with the boys than to call off the voyage half way along. We ended our voyage with an average catch.

In Nova Scotia there are sixty-four shares in a vessel, and there could be as many as fifty or more shareholders, which was the case with the *Vera P. Thornhill*.[7] The shareholders called a meeting, and the majority decided it may be better for all concerned if I would resign. They almost voted to sell the vessel, but that proposal was rejected, and it went fishing another year with a new captain. At the end of the next voyage she landed less fish and was sold to a West Indies company at Barbados.

After we discharged our fish at La Have, I planned to return home, first by rail to North Sydney, to reach the Gulf ferry. But the company asked me to carry the men home in the vessel, clean and paint her and put her in condition for next year's fishing. When these necessary things were done, I was to bring her back to La Have, to be moored for the winter. They offered to put me on wages immediately until my return, and I agreed to do so.

We left La Have when the men were settled and had all their supplies on board. I landed all but five of the crew, and, with the rest, I took the vessel to my home, where we cleaned and painted her, then sailed for La Have. We arrived there on the 15th day of November. It wasn't easy for me to give up my first command. A finer vessel was never built. God knows, I tried as hard as any man that ever went to sea to make a success. I know I made mistakes, and will as long as I live. It was a hard struggle from the beginning of my first trip to the end of the last one. When I left the *Vera P.* I said to myself, "Well, it's not the end of the world yet. I'm going home now, thank God," to marry the teacher that I first met in 1918 when I returned from my first year at the bank fishery. We had been corresponding for the past ten years.

A couple of days before I left for my home, a Captain Rose from Jersey Harbour, Newfoundland, came to me and asked if I would be satisfied for my crew to go with him. He had just purchased a vessel from La Have. If my men helped him take his new vessel home, it would save time and expense by not having to transport his men to Nova Scotia. I mentioned it to my crew and they all agreed.

Now I had . . . to go back to the dory again

I married Ruth Williams on the 26th of November, 1928.[8] Captain Rose and my crew all arrived at their homes a few days later. My wife and I had been married only a few weeks when we made a trip on the coastal boat, and met quite a few captains who were returning to their home at Grand Bank. They had just moored up their vessels for the winter. We chatted about different things and were glad to be in each other's company again. Many thoughts were going through my mind, being one of the captains for two years — what I wanted to be so bad, and worked so hard for. And now I had to knuckle down to the same thing, go back to the dory again, especially after just being married. It wasn't a very good year of the bank fishery. Fair voyages were caught, but the prices were poor. There was a lot of talk about Depression.

One night about one month later, mail came from Lunenburg and I didn't make as much as expected. The owners kept so much back, like crooks. I said to Ruth, "There's no harm to say, 'If I haven't got a downfall!' " But she said, "Arch, you haven't got a downfall. You've learned your lesson. You're just on your feet, and you're going to be skipper again. And the next time you'll make a success of it." I kept that in my mind all the time, and there were times I was so discouraged that I couldn't see how I could. I knuckled down in the dory again. Then that Depression came on.

And I vowed that, next time, I was going to be skipper. If I told a man to get up at one o'clock in the morning, he was getting up. No matter if twenty-odd men said they wouldn't do it, I'd bring her to land! And if I didn't get another crew, I'd lose my skippership again. I was going to be skipper, and I was going to do things my way. Just like I did when I was with other skippermen. Because I never told a man, in my life, since I was skipper, to do anything I hadn't done myself. And

I never put a man over the side of my vessel in any rougher weather than I was put out in myself. What I could do, someone else could do.

Nineteen-twenty-nine came along, and it was getting towards spring when Captain John T. Thornhill, from Grand Bank, the same man who sold me his shares in the *Vera P*. *Thornhill*, asked me to go with him in dory and as mate on the *Paloma*. I told him I'd go in dory, but not be mate. I left home the usual time, around the last week in February, and went to Grand Bank to join the vessel. I had made up my mind. It's just as well to face it. I had to make a living. We left Grand Bank the first week in March and went to the Western Banks.

A monstrous sea . . . cleaned the deck

On our second baiting, fishing in the Gulf some distance off Cape George, St. Georges Bay, we were laying one day in a gale, with just the jumbo and storm sail, when a monstrous sea struck us and cleaned the deck. Every dory (eleven) and everything else on deck, including our deck engine, was carried away. Fortunately, nobody was on deck at the time. Captain Thornhill was looking through the cabin gangway at the time and saved himself by diving into the cabin. The vessel went so far out on her beam ends, our sea-chests turned bottoms up in the cabin, and turned over the salted fish in the vessel's hold. We were left without a dory to save our lives if need be.

When the storm was over, we abandoned our trip and sailed for Grand Bank. We talked about it many times afterwards, how near we all were that time to our doom. We ended our voyage this year with a fairly good catch,[9] but prices were getting worse all the time.

I bought a house in Grand Bank after our last trip that year. It cost $1,100. I had only $800 towards it. I went to Mr. Carr, manager of the company I was sailing with, and asked for a loan of $300. He didn't hesitate at all, although money was beginning to get pretty scarce, and $300 for a man going in dory was a lot to ask for now. He said to me: "Sure, but it will cost you seven per cent. Be only a matter of a year when you gets skipper again and you'll have it paid back. Your prospects are good." That sure gave me some encouragement.

I went home to Anderson's Cove in another vessel, took my wife and belongings, and went back to make Grand Bank our home. After we were settled in our new home, I went to Nova Scotia with Captain Thornhill, the man I sailed with this year, to bring back the *Clara B. Creaser*,[10] a vessel our owners had purchased at Riverport, Nova Scotia. Just one hour after our return to Grand Bank, the 1929 earthquake occurred. We knew nothing of the disaster until the following morning, as the tidal wave did no damage at Grand Bank.[11] 1929 was a very sad year for the people on the Burin Peninsula and in nearby settlements.

She went down like a rock. . . .

Shortly before this tragedy, a number of fishermen were returning from Lunenburg, via North Sydney. It was a custom then to watch for a passage on a vessel that might be coming to Newfoundland with freight, such as coal and flour. A vessel from Pushthrough, Newfoundland, was about to leave and a number of men accepted the opportunity. This was the *Corenzia*, captained by Joshua Matthews of Grand Bank. They left North Sydney late in September, heavily laden with coal. During the night, as the vessel sailed along under full sail, with the captain and all but a three man watch sleeping in their berths, she was struck by a sudden squall of thunder, lightning, and wind. She went down like a rock, with all hands except the three men on watch. Fortunately, one dory on the windward side wasn't lashed, and, just as it was floating from the sinking vessel, the three men jumped in. They had only a few broken paddles and a small bailer to keep the water out, but managed to stay afloat for seventy-two hours in the open dory until picked up in the night by another passing vessel. Most of the men lost were

122 Voyage to the Grand Banks

Before and after the tidal wave struck Burin, 1929

Schooner, believed to be the *Robert Max*, sister ship to the *J.E. Conrad*, c. 1939. She had just enough wind to sail into St. John's harbour, where her dories towed her to anchor. (Photo: Ayre & Sons, Ltd.)

from Garnish, and this cast a dark shadow over it and other communities.

We left in the *Robert Max* for the fishing banks again on the first of March 1930. I was mate now with Captain John T. Thornhill, and my second-youngest brother (Wilson) was in dory with me. He is even smaller than I, so there was plenty of hard work, as usual, ahead of us. We landed our first catch in the last week of April. The price of fish was dropping all the time. Everybody found it difficult to make a living. Labour was only 15¢ and 20¢ per hour, and, to make things worse, money was so scarce that everyone had to get everything charged, which cost a lot more than if you had the cash.[12]

Our voyage ended with a fair catch (2,884 quintals),[13] but we made poor wages. By the time a man with a family paid his expenses and what it took for his family to live on at home, there was not much left to buy food and clothes for the winter months. Things looked so bad for the bank fishery that some merchants were not sure they could carry on next season.

Our first child was born the last of April that year, and,

needless to say, we were proud parents. At the end of the fishing season I went mate on a vessel, freighting. I left again in late October and didn't get home for Christmas. We were freighting from Halifax to St. John's with general cargo, mostly fish to Halifax for shipment to Brazil.

I always went freighting in the fall because it was in my makeup. I didn't want to lay around and get 15¢ or 20¢ an hour, if I could get $35 a month on a vessel. The wages were small, but a fellow has to make a living somehow. I'd go and, when I came home, I'd have something. Anywhere we'd go up there, Halifax, Sydney, Prince Edward Island, things were a lot cheaper than at home. And the customs people all knew that the crew was smuggling in a bit for themselves. The customs were all local people, and they weren't very particular. You'd bring them a little "tip" sometimes. You had to do that. We had to do something, to bring in some meat, a bit of sugar, a few potatoes. You couldn't live if you didn't.

New Year's Day 1931, we arrived at Grand Bank from Halifax after a stormy trip, and within a few days we left again. I ended on this freighting vessel on the 10th of February. This gave me three weeks at home with my family before I left on our fishing voyage. I went with Captain Thornhill on the *Robert Max* again.[14] My brother and I were dorymates.

We found their dory bottom up with no sign of either of them

On our first caplin trip, tragedy struck us while on the fishing ground. We were all away from the mothership hauling in our trawls. As we came on board, we discovered there was no sign of two of the men. I got into a dory with two other men to search for them. We found their dory bottom up with no sign of either of them. The dory was hooked on the trawl, so we concluded that a sea sank their dory. They were both married. It was really sad, and, to make matters worse, one of the missing men had a brother on board. He was heart broken.

1931 was one of the worst years in the history of the deep sea fishery. Fish were very scarce on the Grand Banks. Had the price not been so poor, however, it would have been much better for everyone. And there is no mistake about it, the Depression was upon us. It's the first time in my memory that men from the deepsea fishing had to go home with no food and clothing for their families.

I went mate on the same freighting vessel I went on last fall after we gave up fishing. We left Grand Bank with a cargo of fish on the 20th of October. We didn't get very large wages, but it was still better than laying around doing nothing.

That year we managed to be home for Christmas. Money was pretty scarce, but everybody seemed happy. We had plenty to eat and a house to live in. But our year's earnings were so poor that I only managed to pay our debts, and was not able to pay back any of the loan. I paid only the $21 interest during the past two years.

We left home again shortly after Christmas, and arrived back in time to join the same vessel and captain to go bank fishing again. My brother and I were dorymates, as usual.

The entire banking fleet was out at it again in March, 1932, and everybody finished by the middle of October. Last year was one of the worst years for the bank fishery, but 1932 was even worse. Very few men in the fleet made enough to pay their expenses, which meant they had to go home again with no food and clothing for their families. What I say here goes for all the vessels in the Newfoundland deep sea fishery. Our vessel landed as much, or more, fish than a lot of vessels, but our men made only $130 for their season's work![15] That's not even an average of $20 a month for the seven months of the voyage — from March to October. On our last trip, we fished on the Labrador, brought home 1,800 quintals for our trip, and shared a mere $80 per man. You would have to live a long time to forget a year like that.

One Grand Bank company, Forward and Tibbo Ltd., advanced food for the families of all the men coming back with them again the next season — and that was probably all of them. This company had two vessels, with a total of fifty men. Another illustration: After I settled up, I had ten dollars coming to me after my expenses were paid, including the $21 interest on my $300 loan. Only one child in our family, so it's easy to understand how people with large families were situated.[16]

It's a serious situation all around

I was thinking seriously about getting out of the fishery altogether for the time being. I didn't have the least idea what I would do. Owing to this Depression, some deep sea fishing companies had gone broke, and it's a serious situation all around. At the end of our voyage, the captain and owners decided our vessel could go on the north side of Fortune Bay and cut a load of wood for fuel. It was impossible to buy enough coal for the winter months because of the bad times, so there was no trouble getting a crew. Eight men, some who fished in the vessel the past season, gladly volunteered for the trip. Our vessel was 170 tons. Every man took his own food and axes. Everybody liked the idea, including the owners. Freights were scarce, and there wouldn't be anything for the vessel to do. And there wasn't any employment for men, except for a small number working in the fish store, packing fish in barrels, etc.

We went to Bay D'Espoir and moored up. It was a new adventure, the first time a banking vessel of this size was known to be used to go to the woods to make a living. Our trip took eight days from Grand Bank. The vessel's hold was loaded and she was piled up on deck. When the wood was discharged, it was measured by the cord. The owners took one-half and the other half was shared amongst the eight men. By the time this was finished, it was getting along into November. Then the owners hired five men, including myself, for 20 cents per hour to paint the vessel and carry her to a safe harbour where she was moored for the winter.

I was looking around and wondering what I could do next year, and the way things looked, if I found anything at all, it would be as good as fishing. On the 20th of November the manager of the company I had been serving for the past

Banks schooners at winter moorings, c. 1920s.

four years sent for me. He asked me if I would take one of their vessels to Prince Edward Island for a load of produce. I gladly accepted it, and who knew what it might lead to? God knows, we needed a few extra dollars badly. For the past few years it had been, and still was, a struggle to make a living, and nobody knew how long it was going to last.

The following day I left Grand Bank with five men, bound for Prince Edward Island. The manager made it clear to me that they needed me as captain only for one trip. They had made arrangements for another man to go as her regular captain. He was investing in the vessel. We made a quick trip to and from P.E.I., and all the crew got their produce and other things for their winter at a much lower price than would be paid at Grand Bank.

Besides some produce, I brought back some chickens and fresh pork for my family. That pork was good pork, fed on milk. A farmer came aboard one day and sold me a hog that weighed 103 lbs. for three dollars. The company brought down quite a bit of fresh beef, and fresh pork, and probably paid less for it there than I did because they bought it in bulk. They paid, maybe, one cent per pound duty on it. But it was sold in Grand Bank for 15¢ per pound or more, up to 20¢, never cheaper.

I arrived on December 8, at nine o'clock in the evening, and at 11 o'clock our second little girl was born. My wife and I were overjoyed. After having the cargo discharged, the vessel was moored up until after Christmas. I was very pleased to have the few extra dollars I just made for Christmas.

It was about the last week in December of this year, 1932, when I accepted a call from a man to go mate with him on a vessel called the *R.L. Borden*, owned by Grand Bank Fisheries, the same company I had been with since I left the *Vera P. Thornhill*. The mate's wages were $35 a month; the captain's, $75; and the crew's, $30. At least we knew what we were making and what we could spend. Our vessel was one of the largest and had been banking for several years. But, owing to the Depression, the owners decided a few years ago to take her from the fishery and send her freighting.

Chapter Five

Barbados She's Goin'?

Since confederation with Canada at mid century highroads, trucking, and airline development rapidly changed Newfoundland communication, transportation, and employment. It may be easy to forget that from its earliest European settlement maritime shipping linked Newfoundland's communities with the outside world. Normally ice-free south and east coast ports had a year round maritime trade involving contacts with the U.S.A., and Canadian, European, Caribbean, West Indies, and South American ports. This trade had its small category of specialized merchant seafarers. There was a thin line between fishing and freighting-shipping employment. Shipping employment opportunities existed, as at larger Fortune Bay fishing and export trade ports like Belleoram, Fortune, and Grand Bank, but they were limited. Most Newfoundland schooners and even tern (three mast) cargo vessels, carried no more than six or seven hands when freighting. Arch's story reveals, however, that many Fortune Bay fishermen moved between the two activities either seasonally or occasionally. And some were exclusively identified with 'coasting' and 'went foreign' as work required.

Despite a deep history of marine disasters in Newfoundland waters, Newfoundland skippers with fishing and coasting experience regularly sailed their vessels without serious mishap to familiar eastern Canadian ports, like North Sydney, Halifax, Campbellton, and Prince Edward Island, and to the "Boston States". Most of these men had learned basic seamanship and navigation skills in the informal classroom of shore and/or banks fishing. But not all were qualified to navigate to more distant foreign ports.

One such man was master of the coasting schooner *R.L. Borden*. As we learn here, in the winter 1933, when he asked Arch, his mate, if he "had navigation" adequate to take them to Barbados both faced a dilemma. Depression hardship had battered the fishing industry and fisher earnings, and Arch looked forward to stable monthly 'wages', even if only 'on account' with the vessel's owner merchant. But his skipper would have to replace him if he didn't have the requisite navigational skill. Their joint solution to the problem illustrates a blithe self confidence that is perhaps essential to successful seafaring. Here it is girded by religious faith.

By 1934 Arch had spent four more hard years back 'in dory' when offered the chance to skipper a fishing vessel again. The schooner *James and Stanley* was owned in Newfoundland. He tells us he returned to command with a resolve to abandon 'counting fish'. What had changed his thinking about this labour incentive scheme since 1927? His reasons sum up as operational efficiency, safety, and crew cooperation. But, now a family man, more seasoned, and compelled to consider abandoning fishing altogether because of the poor economic conditions, did he better understand how financial uncertainty damaged fishermen?

He knew his role was insecure. His personal future and his crew's earnings depended upon his vessel's success. The responsibility is complex. To fulfill it one has to "have the push in you", as Arch argues. But success involves the skipper's every resource; experience, judgement, the ability to find fish, and more. Every fishing decision involves uncertainty and the indifferent sea tolerates few mistakes. Loss of the Grand Bank schooners *Alsation*, in 1935, and *Partanna*, in 1936, and their entire crews, are ikons of the risks that shadowed their work. Little wonder that these skippers worried a lot, about fish, their crew, and making it home again.

1933-1938

New Year's Day 1933, I signed articles as mate for a twelve month stretch on the *R.L. Borden*, with Captain Charlie Rose. There were plenty of cold days ahead of us. We went freighting all that winter to the usual places, St. John's, Halifax, and P.E.I. In the first week in January we sailed from Grand Bank en route to Halifax with a load of fish, and took on cargo for Grand Bank. We kept doing this until the middle of February, then we went to St. John's with a load of oil. The weather wasn't surprising to us — plenty of gales and frost all the time, in the winter season.

The last of February we loaded again for St. John's. We encountered heavy Arctic ice during our trip and were delayed at Trepassey for eighteen days. We climbed the hills above the town and watched the sealers, a great number of them, drifting up towards Cape Race. By the time we reached St. John's, a number of them were there with full loads. After being there for a week, another cargo was taken on board and we sailed for Grand Bank again. We kept freighting loads of (likely, fuel-) oil back and forth from St. John's to Grand Bank.

We arrived at Grand Bank in the middle of April from one of our trips, and the manager called Captain Rose to his office and asked him if I had navigation. Captain Rose was a deep sea fishing captain from his early years, and this was his first year absent from it. A Newfoundland fishing skipper needed very little navigation, but a lot of judgement and experience. All fishing vessels on the Grand Banks used magnetic charts, and all variation was on the charts. That's all you need to know to take off a course. The manager explained to the captain why he asked him if I had navigation. Knowing the captain did not have it, Mr. Carr said, "we're going to send you to Campbellton, New Brunswick, to load lumber for Barbados, and bring back a load of molasses to Burgeo, Fortune and Grand Bank. If the mate haven't got navigation,

Grand Bank, Newfoundland, early 1930s. "Women resting on beach following task of turning drying cod."

we'll have to select another mate who have."[1] We were discharging some freight when the skipper came on board and asked, "Arch, have you got navigation?" I thought it strange for him to ask me that, and I said, "No, sir. The only bit of navigation I have is four lines, 89 48, read your sextant, and whatever that reads, add it up or subtract it, what it might be, and get your declination out of the Almanac." That gave you your latitude and longitude. That's all I knew. Then he told me the story. He said, "As you know, I got no navigation." Now, he couldn't even take a latitude sight. He never even had a sextant. But I didn't have any more navigation than he did, with the exception of knowing how to read a sextant. My cousins taught me that one time. I began to think quick; one of the last things I wanted to happen was lose my job. I was getting $35 a month, the captain and I got along well, we lived next door to each other, and I knew he wasn't in a hurry to lose me either.

He said he'd get us there, so he must have navigation

"Well, sir," I said, "if you're so game as me, you say to Mr. Carr, 'Yes, Arch got navigation. He said, "Barbados she's going? I'll get her there."' If you want to tell Mr. Carr that, I'm game to do the rest of it." The old skipper was pleased to hear me say that. "Oh," he said, "I don't give a damn. That's alright." When he went up in the office, and Mr. Carr asked, "How did you get along with Arch?", the skipper said, "Oh, he got navigation, I suppose. He said he'd get us there, so he must have navigation." Mr. Carr was pleased and said, "That's good enough." That's all they bothered.

My wife had five brothers, and four of them were foreign-going captains. Two of them, Captain Clar Williams,[2] and Gordon, were in Grand Bank at the time. One lived there. I went on board Captain Clar's vessel and told him the story. He said, "What do you want me to do?" I said, "I want you to come up to my house tonight, show me a few lines, teach me navigation as quick as you can." He agreed he would. He had two more days in Grand Bank before sailing and we had the same.

How thick is your bloody head?!

Around eight o'clock in the evening he came to our house with his books, one almost too heavy to carry. We knuckled down to business. Believe me, he wasn't an easy teacher by any means. The first thing that would go wrong, he would say, "How thick is your bloody head!?" "Oh, boy," I thought to myself, "what am I up against?" He stayed until midnight, and I kept at my books until two o'clock in the morning. And my wife, she was pretty good, because she grew up with her brothers and they used to teach one another in their own home, and she was teaching for ten years.

When Clar went on down to his house, she'd get behind my back and drill me in the sixties, degrees and minutes, and all this and that, that I got fooled up on. But I made myself believe I was doing pretty good. And what I couldn't understand, well, she'd point it out to me. So I was doing good. We were at it 'til about 2 or 3 o'clock in the morning, and then I went to bed, had a few hours rest, and went down aboard the schooner, working again.

Next night, Clar came up again at the same time and stayed until midnight. If everything didn't go just like he wanted it, he'd say, "How thick is your bloody head, boy!?" Fun now, afterwards, but that was alright. I felt I was getting places and he was pleased enough when he left the house that he said, "If you can't get to Barbados and back now, it's not my fault."

My wife and I did the same thing again until two o'clock in the morning. We were going to sail the next evening or the next morning. But we were going to St. John's again before we headed up to Campbellton. So I had everything. Anyway, I made myself believe it. I made a big pad. It had a sum, and sine, co-sine and declination, and tides and latitude. My Lord, when it was made up it was half a mile long. But I was well satisfied. "By gee whiz," I said to myself, "I've done pretty good." But there was one thing I could not grasp and I didn't like to ask him too much.

I went on board Clar's vessel the next morning, before he was to sail. When I went down Clar was there. "How did you get along afterwards, b'y?", he asked. I said, "Oh, I think everything is alright." Now the one thing I couldn't catch was the variation on the true chart, and I didn't want to tell Clar because I knew what he was going to say. He'd call me stupid again. But he was a good man, and we got along well ever since we first met.

It seemed as if I felt it strike my brain

His brother, Gordon, was a foreign-going captain at one time, but he was cook with Clar now, so I went to Gord'. I went down in the forecastle, and he said, "How you getting on, b'y? Last night I hear you're doing pretty good." "But, Gord'," I said, "there's one thing, boy, I couldn't seem to catch on to." And he said, "What's that?" "Well, the variation on the true charts. I don't know, I can't." "Come on," he said, "let's go back in the cabin." So we went back in the cabin of the schooner, because he knew just as much as Clar did. He took the true chart out, not the magnetic one, and started to show me about the variations, and it seemed as if I felt it strike my brain. I caught on just as quick as that. Within the hour they sailed, and we sailed early the following morning.

Since we were to make a few more trips elsewhere before going to Campbellton to load lumber, I had ample time to do some practising. And I needed plenty. Our first trip after leaving Grand Bank was to St. John's, for a load of oil. While there I went ashore, we got our compass adjusted, and rented a chronometer for our Barbados trip. The chronometer was new to me, and a very important instrument. I got all the instructions from Mr. Sam LaFosse, the person we were renting it from. He warned me to be very careful with it, and later I learned why.

LaFosse put the chronometer in shape and told me I had to wind it up every twenty-four hours. But I forgot to ask him which way you wind the chronometer, and I was out now, up around Cape Spear, heading for Grand Bank. I took a chance and did it right. If it was only a second or two seconds off, it would put you out of position a lot.

Our oil cargo was discharged at Grand Bank, and from there we left for Campbellton, New Brunswick, on the first leg of our new adventure, sailing, no engine. Of course, I wasn't to do any navigating before we were loaded and sailed off the magnetic chart, which would be over 100 miles out to sea. I had all the necessary books and was practising all the time. Sometimes I'd make mistakes and I'd dig it out and get it right. I knew right where we were because we could see land. Plenty of mistakes, but I said to myself, "I'm doing pretty good."

We were at Campbellton approximately ten days, loading shingles, and sailed on a Saturday morning. We came out through the gut of Canso at night. Next morning, a Sunday morning, we were out through, sailing along just beautifully.

At 7:30 a.m. the sun was shining, and it was a good time to give myself the first test by taking an altitude reading. Everything seemed fine. We were probably fifty to sixty miles off then, and couldn't see any land. I completed the sum, and, believe me, it wasn't a short sum either, and I took my position off the chart, and it had us more than 100 miles in over the land! My blessed Lord! In all my practising on the last trip coming to New Brunswick, I never made a blunder like this!

I kept cool and never said a word to anybody. . . .

I kept cool and never said a word to anybody on board about it, because I knew the skipper'd be worried. Just simply took my time to find the stupid error, got my book out, and I thought, and thought, and thought. It was half an hour before I found that simple mistake. I knew then what it was. I was alright. I took my seven o'clock reading, twelve o'clock latitude, and four o'clock sight and got my position on the dot. I knew because we were towing a taffrail log — always used it in local sailing. The skipper was in charge then until we got off the magnetic chart, which would be fifty to sixty miles outside Sable Island. We went on, and on, and on, the skipper doing fine on his magnetic chart. The next day I took a sight again and it was pretty good. So, next evening, when we got outside, off the magnetic chart, the skipper said, "O.K., Arch, she's all yours now." Of course I already knew that.

A funny feeling went through my veins, but I didn't mind it. I had as much confidence in myself as any man that ever sailed to sea. I knew it was going to be quite an experience sailing to a little rock, a little speck on the chart, over 2,000 miles from our point of bearing, and the nearest land to it is around 100 miles. But I was thrilled with the new adventure.

I said to the fellow who was on watch, "Alright, pull in the log, boy." I was sitting down in the cabin talking to the skipper when he passed the log down to the cabin for me to read. He thought I was going to read it and put it overboard again, but I never looked at it. I said, "Put it behind the belayin' pin there. I don't want that." The skipper looked at me and he said, "Arch, you don't mean to say you're not gonna tow the log down to Barbados?" "No, sir," I said, "if I make this, I want to make it fair play, not foul play. I'm not towing no log." My Lord, he didn't know what to make of it.

After a few days sailing and reaching the Gulf Stream, we struck bad weather. Heavy rain and strong wind. We hove to under the foresail, and had the others all snugged away. We were hove to for forty-eight hours, and I never saw the sun, didn't get an altitude for over three days. I was just using my dead reckoning now. When the storm cleared away and the sun began to shine, I began to shine myself. I got a beautiful sight at seven o'clock in the morning. I might have been seven or eight miles out, but I put us on course and took us to the south, straight as a line. Every day now the weather got better and I got my regular altitudes three times a day. We were now eight days on our journey and what a lovely trip we were having. We were in the easterly trade winds all the time now and the sea was so smooth as if sailing along on a pond. Words can't express how much I enjoyed it.

I had the confidence...and a lot of faith in prayer

But I think our captain was a bit nervous sometimes. He said a few times that he couldn't understand why I wasn't towing the taffrail log. I said, "Sir, if I find Barbados, I want to find it fair play." I knew I was going to find it from the beginning to the end. I had the confidence in myself and a lot of faith in prayer.

After sailing for twelve days and nights, we were then about 300 miles from Barbados, and the captain came to me in the evening. He said, "Arch, if we sees a ship tomorrow, we'll signal her for to get our position." "Well, sir," I said, "I don't think there's any need for it, but if you feel that way, you're the skipper." But I had enough confidence in myself, I didn't want to ask for a position. "I think we will," he said. "After all, you only had two nights to learn navigation." I said, "Yes, sir, but I had a good teacher."

By God, the next morning, around eleven o'clock, there she comes: a big four stacker. A big liner hove in sight and seemed to be coming in our track. The skipper said, "Alright, put the flag up. She'll come alongside." So we ran our flag up, and I'm praying to God he wasn't going to notice us, and she went and went. That's what happened; the ship passed within a mile of us, but they never came closer. He went right on and never noticed us. He used to swear, the skipper did. "You know," he said, "that S.O.B., he's not gonna come down to us." "No," I said, "it's too bad, isn't it? But we'll take a chance and go on." I pitied the poor old skipper in a way, but I got my wish.

We went on and were getting pretty well down now. Our 14th day, at seven o'clock in the morning, I took my altitude as usual. Another fellow and I went off watch at 8 o'clock, and usually go below. When I had my sight made up, I said to the skipper and crew, "If we don't see Barbados at 11 o'clock today, I'll eat my shirt." Oh, boy, many happy faces, including the skipper's.

Get up, boy, and look at Barbados land

It was very warm then and my watch mate didn't bother about going below. He went to sleep on the shingles we had bulked up on the forward part of the deck. It's a God's fact: at quarter to eleven I was walking along on the shingles, and I went along and kicked him in the backside and woke him up. I said, "Get up, boy, and look at Barbados land." The weather in that area was hazy, no fog, and you can't see any more than three or four miles. But, nevertheless, after sailing fourteen days, a distance of 2,000 miles, Barbados was so fair by the top of our bowsprit as if on a line that was drawn with a pencil mark. How happy I was. We went on, went on sailing. We had a cook on board and he went down there a good many times before. He didn't have any navigation or anything, but he said, "By God, I've been down here a good many times and I don't think that's Barbados." Now the nearest land to Barbados is eighty miles — Turk's Island.

The skipper said to the cook, "What, boy? You don't think that's Barbados?" "No, sir, I don't think." "Well," I said, "there's no trouble to find out. There are some fishermen there in those little boats. Go along, speak to the fishermen." We came up alongside this fisherman and "Hey," said the skipper, "what land is this?" "Oh, man," he said, "Barbados. You can keep right-a-way, now, and go right in the gut."

We were in and tied up a short time after taking a pilot. After getting through the customs, discharging was started immediately. The crew started, but the weather was so hot, a crew of local people had to be hired on. After our lumber was discharged, a cargo of molasses was taken on board. We enjoyed every minute of our stay at Barbados. It was the crew's first time there, except for the cook. I taught him navigation while on our trip. All the way back across the Atlantic again we enjoyed another pleasant passage. Coming back was the same thing. We were exactly the same number of days as we were going. I made it so straight for our first stop, Burgeo, as if you'd drawn a line.

I was told by the manager before we left Grand Bank that my wages would be increased while on this particular trip. We were at Burgeo only one day discharging the part cargo for that place. We left just before dark and arrived at Grand Bank early the next morning. I now had enough confidence in myself to sail anywhere across the Atlantic.

We were kept busy the remainder of the year. The banking vessels ended their voyage as usual around the first week of October. The quantity of fish had been fair, but prices rotten again. Our little wages here were about equal to what the men shared fishing, which was not very much. A week before Christmas we left Grand Bank bound for Halifax. As we were about to leave I was called in to the manager's office. He asked me if I would consider going master of the *James and Stanley*, one of their large fishing vessels, in the coming season. Needless to say, I accepted and felt pretty happy over it. I had already been fishing in the dory with another man for four years, and one year freighting as mate, so I was glad to get back as captain again.

Money could never pay for what that trip meant to me....

That Christmas we spent at Halifax, and arrived home the last day of the old year. On New Year's Day, 1934, I signed off articles just one year from the day I signed on. A few days later, I went to the office and settled up for my year's work. The account included $21 interest, but I didn't see any mention of the extra wages for the Barbados trip, so I asked the secretary, "What about that trip when I navigated the vessel to Barbados?" He said, "You'll have to see the boss about that." I did, and he looked out through his door and said, "Give Arch an extra five dollars for that West Indies trip." The money was very small, but money could never pay for what that trip meant to me and never will.

I was home with my family for a couple of weeks before I took over the *James and Stanley* for the first of what were three hard, struggling years. Quite a bit of work had to be done

Bowsprit schooner bound for the banks. circa 1930s. Skipper Arch believed it to be the *James and Stanley*, just leaving Grand Bank gut.

during the five or six weeks before fishing began. A new mast was put in her, and on February 20th I sailed her around to Burin for docking. She came off dock eight days later in time to take on our men waiting at Grand Bank. I had a full crew, most of them from Fortune Bay.

The men would cuss this counting fish

Earlier I mentioned my younger cousin, Frank Thornhill, when we were dorymates in the *Dorothy Melita*, eight years before. This year he too was a new captain with this company. He skippered the *Laverna*. This year meant a big change for us in another way. Frank and I had vowed to each other years ago, that if the day ever came for us to become captain, and we felt it would, on the first trip out, we'd throw the tallyboards overboard. We wouldn't count another fish!

That's all we ever talked about when we were fishing. The men would cuss this counting fish. Before they left this world, some poor fellows said they'd never get any forgiveness for how much they cursed counting fish! It was the biggest cutthroat thing. Sometimes it was so bad that men aboard the same schooner went for days and trips without speaking to each other. They'd sit at the same table, looking at each other, and yet cutting each other's throat. I've seen men come aboard in the daytime, and if you got more fish than the other fellow and went to throw your painter up aboard for him to haul you alongside, the other man wouldn't take it. He'd walk away.

He'd done so badly, only landed 200 fish — but large fish, all his dory could carry, while the other got 1,000 small, from the same day's work. Now, your large fish was thrown in with his small ones, and then both have to do the same work, pitch in, put down and salt all 1,200 fish, while the one man's large fish are paying for the other's snappers!

I've seen times when men stood all night at the same splitting table, dressing the fish, and never spoke to one another. There was a reason for it. The man that landed the most fish in weight didn't get paid as much as the fellow with the most fish in the count. What you earn is made up from the count. And the man that wasn't getting any fish worked a lot harder than the man that was, because he'd have to haul his gear up, put it in trawl tubs and move it somewhere else, sometimes a couple of times a day. While the other fellow a berth away is just hauling his gear back and forth across his dory, bringing plenty of fish aboard.

Then you'd go aboard at night and the deck would be full of the other man's fish, and you'd both have to work the same to put it away. But you never made a dollar that day, while he made $25, and maybe you have a bigger family to support. That would go on for days, trips.

I know good men who ended up low dory

I know good men who ended up low dory. We used to say then, that there are only four men, two dories, they'd bother about: the high dory and the low dory. When you'd come in off a trip, the first thing the owner would ask the skipper was, "Who's high dory?" Because, natural enough, some day the high dory skipper is going to get to be skipper of a schooner. And many did.

A lot of skippers, and probably the owners too, had the foolish idea that some men wouldn't try. Without a doubt, a few would say, "The hell with it. If we don't get it, somebody else will." They had the idea that you'd get more fish by counting, but I never believed it. You lost a lot too. And the men knew it. They were overjoyed when we threw the tallyboard over, and within less than two years, not a fish was counted in Newfoundland.[3]

We had a fair voyage this year, just enough for them to say, "Well, we'll give him another chance." Our frozen baiting began on March 15th and the spring trip ended around the first of June. It was a good trip, but other vessels had more. Not much exciting happened during the season that ended on the tenth of October.

My cousin, Skipper Frank, did much better. When he found the fishing poor on the Grand Banks on our caplin trips, he took his vessel to the Labrador with cod jiggers, which wasn't done very often. It proved a lucky move this year, as fish were numerous up there. He went as far as Cape Harrison, where he found fishermen using cod traps were getting more fish than they could handle. The two captains agreed to work together. The bank fishing skipper put some of his men on board the cod trap skipper's vessel to help split fish, and the bank fishermen hauled the cod traps.[4] Both vessels ended up with bumper loads. After discharging her Labrador fish the *Laverna* went to the banks again with further success. The price of fish did not improve very much, but most of the deep sea fishermen had sufficient food for their families in the coming winter. My men were quite happy with the year's work. After landing them, we went back to Grand Bank, cleaned and painted our vessel, and carried her up Fortune Bay on the 15th of November, and moored for the winter.

You had to have the grit in you, the guts. . . .

Most of my men came back with me in the spring. As a rule, there's always some change each year, probably five or six men. I have to say that I enjoyed my season's work. By this time I had learned that you had to have a push in you, that you couldn't take everything easy. A skipper is responsible for twenty-odd men, twenty-odd families, his own family, and then the owners. It's a big responsibility. But you just have to have something behind you.

    I know men who made the best of mates — second hands. But when it came to being skipper, they were no good, never got enough fish to eat. They didn't have the push. They were too easy. You can't be too easy when you have a crowd of men. You've got to make your living. There are times when you think you were pushing them hard, but I never told a man to do one thing I wasn't told to do myself. And I figured, if I could do it, by jingoes, brother, you have to do it. You had to have the grit in you, the guts and everything else. Some say you got to be the devil. If you're called a "good" man when you're skipper, you're no good. I don't see any possibility that you'll make a success of it if you are called a good man.

Once a man came aboard my schooner and he was telling us about this particular man he was with. And he said, "Oh, skipper, he was a good man. What a lovely man he was. Sir, if there was any work to do up aloft on the masthead, he wouldn't allow we to go up. He'd go up his ownself and do the work and say, "What odds, boy, you don't want to go up there. I'll go up there." "And," he said, "when we wasn't fishing he'd come and get up in our bunks and drive works (skylark) with us. But he was some man, skipper." Well, that man never got enough fish to eat while he was fishing.

I handled men differently now. You'd get some good men and some bad. I remember one fellow. I had him sized up. He was up in his bunk pretending to be sick, and the rest of the men were out in their dories. I heard about him, went down in the focs'l and said, "Brother, if you don't get out of this bunk, I might look pretty small, but I'll have you yanked out. You won't be hurting. You'll be sick." No man was getting off with that. I know when a man is sick. Brother, he wasn't long getting out and into the dory. Never got sick afterwards.

1934 had been a lot different for me compared to the last five, and I felt on top of the world again. But there are many setbacks in life, they could come again, and I had a long way to go before I could feel secure. I looked forward to spending winter at home with my family. I had been at sea, summer and winter, for the past three years.

She never came, first nor last

There were about sixteen deep sea fishing vessels sailing from Grand Bank in 1935. It was still a lively spot, but there were fewer three-masted foreign-going vessels than in earlier years. The first of March, as usual, the year started off with plenty of activity in Grand Bank harbour, as everyone got ready for the fishery. By the tenth or twelfth of March, most vessels were on their way to the banks, out on their frozen baitings, and expected to arrive back again sometime in April. When the time came they began to arrive, one after the other. On some days only one arrived, and most days three or four, until they were all in. But this time there was one that never came.

We were the eighth or ninth vessel to return. As a rule, all the Grand Bank vessels saw or spoke to each other some time on the trip — there were no ship-to-ship radios then. When the first vessel arrived, people always gathered around to get all the news, and the skipper and crew would be able to say they saw and spoke to quite a few vessels on their trip. This went on until all the vessels arrived.

Many people were on the pier when we came in, and the first thing they asked was, "Did you see any sign of the *Alsation*?" Because she was the only one that had not been seen or spoken to by the vessels that had arrived. She never came, first nor last. The day she left her home port, at about the time she was getting off from land, one of the worst winter storms in years came on. Not a thing was ever seen of the vessel again.[5] Twenty-five lives were lost on the *Alsation* alone, owned by J.B. Patten and Sons. And another vessel on her way from Fortune Bay to Gloucester, with a load of herring, the *Arthur D. Story*, was lost in the same storm with all hands.[6] It cast a dark shadow over Grand Bank and many other places where husbands, sons and fathers belong.

Most vessels had good catches on their first trip, and I was very pleased with ours. Everyone hoped that there would be good news from the *Alsation* by the time we returned from our second baiting, but it never came. As the year progressed, we did well, trip after trip, and ended up second to the highest vessel. Our first son was born to us on September 3rd, and we were proud parents when he arrived. The price of fish was still very low and, there's no mistake about it, people had to struggle to make a living.

1936 was a year of hard struggle from beginning to end, and full of sadness for many people. As usual, Grand Bank harbour was cleared out within two or three days when all the vessels went on their first baiting. But we found fish very scarce. The forecast was good and the barometer steady one evening towards the end of our baiting, so I decided to set our gear and leave it in the water all night. I gave the men orders to set twenty-six lines per dory, and they arrived on board the mothership just before dark. It was as nice an evening as ever seen, and we were all counting on another two or three good fishing days in this spot so that we might end up with a good baiting.

We were running before the seas. . . .

But at about two or three o'clock in the morning, a storm came on. All of the men were called out of their berths to slack out all the cable we had, so the vessel wouldn't go adrift. We were anchored in shallow water at the time, only about nineteen fathoms. By ten o'clock in the morning, the storm was so severe I thought we'd sink at anchor, so I ordered the cable cut, to run before the gale. We were running before the seas when a bad one struck us, broke our foreboom and tore our foresail to threads. But we kept running under bare poles until the storm abated some time during the night. By then we were over 100 miles from the gear we set the evening before. Our trip had to be called off and we headed for our home port.

What a mess we were in: eleven dories' gear — over 300 lines, besides the anchors and buoys, were completely lost, foresail torn to shreds, booms broken, 250 fathoms of eight-inch cable gone, an 800-pound anchor lost, our lighting plant hit out of commission, and only 200 quintals of fish for our baiting.

When we reached Grand Bank there were quite a few vessels there that had lost all their gear in the same storm, but they had much more fish than we. I was never more discouraged than I was on this particular trip, and the owners gave me a poor reception. But I guess that was to be expected.

After a couple of days, we were fitted out and sailed again, taking on another baiting. We were just about the last vessel to leave Grand Bank; and one of our vessels, the *Partanna*, hadn't arrived yet. Nobody had seen her since she sailed on her frozen baitings, and everyone wondered if it was going to be another tragedy like last spring. She was given up for lost with all hands — twenty-five men — by the time we returned to land for our second baiting. She was last seen leaving Harbour Breton, headed out to sea between St. Pierre and Miquelon.

One of the worst storms that year came on the same day, but we never knew when she was lost. Maybe he had fished his frozen baiting on the Grand Banks, although not many went there in March month. I think he did, and ran up on some shoals around Cape St. Mary's or Cape Race. Only some of her dories were found. They drove ashore in the Cape Race area, and a fellow at Baine Harbour, Placentia Bay, picked up a piece of wreckage with the last three or four letters of her name — *anna*. He nailed it on his store where it stayed for years. That was about forty-eight or fifty lives lost from these two vessels in two years. Pretty sad for a small place like Grand Bank. And within a short time of this, one of the three-mast vessels coming from Portugal in the winter was lost with its entire coasting crew of six hands, all from Grand Bank.

Our spring trip ended with a fair trip, but not so much as we'd like it to be. After landing our caplin trip, a supply of bait was secured and most vessels went to the Labrador for the fall trip. On the 9th of September, my cousin, Captain Frank, lost the *Laverna*, when she drove ashore in a gale at Salmon Bight Passage.[7] We were all fishing in the same vicinity when she grounded on a reef, could not be refloated, and was a total loss. Within a few days, the *Paloma*, another of our vessels fishing in the same area, caught fire and was a total loss.[8] Our company had now lost three vessels in one season!

The owners didn't give me a very good reception

We fished until the first week in October, when the weather

turned bad. Since our vessel was one of the largest, and we did not have a big trip, there was plenty of space in the hold. Quite a few of the stationary fishermen fishing in this vicinity were from Conception Bay, and asked me if I would carry them and their fish home on my way. I thought the freight and passage money would help us make a few extra dollars, especially with a lean trip, so I agreed. And I felt sure the owners would appreciate it.

We loaded our vessel with some forty passengers, their fish and belongings, and quite a bit of wreckage salvaged from our two vessels. Then we took on one man's seventy or eighty barrels of cod oil for a deckload. So we had quite a cargo, just about a full load of fish in the hold and a big weight on deck.

Just before dark, we left Salmon Bight, Labrador, and sailed down the Labrador coast via Belle Isle Strait and headed our course outside the Funk Islands. The next night, we were about thirty miles northeast of the Funk Islands when a severe storm came on. We hove to with just our foresail on. Our vessel was leaking and it took all of our time to keep up with the water. Every one of our forty passengers and twenty-odd men took his turn at the pumps. By the third day out the water was gaining on us and we could see it from the cabin floor. It looked serious.

I told the men to take the iron pin mauls, bash in the barrel tops to let the oil out and jettison them in order to save our lives and the vessel. The barrels weren't long going over the side. When the weight was removed from the deck, we could keep the water out. The storm abated on the fourth day, so we could increase our sail and make headway. But not for long. We had to shorten sail again within hours, and were almost eight days before we reached Conception Bay.[9] It wasn't until the first of November that our men were all landed at their homes, so we were late getting home this year.

Although the season ended with another fair voyage, the owners didn't give me a very good reception. It would have been alright if we hadn't jettisoned that man's deck cargo.

The manager said I shouldn't have taken the extra fish or passengers on board. But I thought an extra $700 would be a good help, especially on the end of the voyage. I expected a court case about the jettisoned cargo, but that was the least of my worries. I did it for the well-being of all concerned.

I saw tears in men's eyes. . . .

I had been home in Grand Bank for a few days after landing my men that fall, when another Grand Bank company, J.B. Patten & Sons, asked me to be captain of one of their fishing vessels the next season. It was the *J.E. Conrad*, a large, modern knockabout of 170 tons. They had bought her down at La Have. I hesitated only briefly before making up my mind, as the vessel I'd been in for the last three years was getting along now. But I liked my company, especially the manager. His equal never stood in shoe leather. I went to him, told him about the offer and the vessel's name. He replied, "Skipper, it makes all the difference in the world, the kind of vessel you have. And I'm sure you'll do well in her." He asked me to stand by them until the *James and Stanley* was ship-shape for the coming season, and who I could recommend to go as her new captain. I recommended one of my first cousins, who was second hand with me in the *James and Stanley* for two years.

I had worked on the *Stanley* for only three or four days when the *J.E. Conrad* arrived in Grand Bank. Her owners called me and asked if I would take her over right away, land the men at their homes, and then get her ready for the next voyage. I went to my boss and told him what I was asked to do, and he sure understood.

I asked myself, "What in the name of God have I done?"

The *J.E. Conrad*'s crew had just made a very poor voyage, and I'll never forget how, of all her twenty-four men, there was only one sack of flour to be landed to their families. I saw tears from men's eyes when they told me about it on the way

Skipper Arch in centre, with brother Eli James Thornhill, seated left. They were being served tea, at Anderson's Cove, c. 1937. He had just moored the *J.E. Conrad* for the winter at Recontre East.

to their homes. It had been tough going for the past five or six years owing to the price of fish. After the vessel was in order, we carried her to Rencontre East, and moored her up for the winter.

1935 and 1936 will not be forgotten for a long time, because so many vessels and lives were lost from Grand Bank. And, owing to the Depression, these vessels were never replaced.

The *J.E. Conrad* was a fine vessel and I liked her very much. I'd admired her ever since I first saw her at La Have, the first year I joined the *Vera P. Thornhill*. My new company asked me to go in the *J.E. Conrad* because of the fair voyage I had that last year in the *James and Stanley*, and now I had a good vessel and an excellent crew of men. But I worried a lot on our first trip in 1937. I'll never forget it. When we came in off our first frozen baiting, we had only 230 quintals, while all the other vessels had 600 and 700. They had enough to wash out their fish and stay in port for a few days, but we didn't. We took on supplies and another baiting, went out again, and caught only enough to make up 650 quintals on two baitings. Not as much as some did on one.

We washed that out and went on the banks for another baiting and only got another 300 on that, while others got 600-700 quintals again. I asked myself, "What in the name of God have I done?" I have a good crew of men, a good schooner, and I was doing exactly the same, working harder than the men that were getting the fish, because they didn't have any worries.

We came in and took another baiting, went out, and I saw another schooner and spoke to him. It was one of the real Grand Bank fish dogs. I asked him how the fishing was. He was just leaving for land, his spring trip finished. We had landed 650 quintals, and had another 300 aboard — 950 or 1,000 total — while that man had over 2,000!

We were up with the top highliners!

I decided to anchor there and we fished. The first day we caught 200 quintals, and, without ever weighing anchor, I got 800 in that same place. So now we had a total of 1,750. Still, I knew others had 300-400 more. It was getting up to the caplin season then, and we had to get ready for our next baiting, so I left for home. The other schooners were in Grand Bank, taking out their fish, but I changed my mind, running along by Cape Race, and ran up to St. Mary's Bay and got enough bait for three days.

Our salt was used up by then, so I collected enough from different shore fishermen, and we went out and got 300 quintals more. When landed and made, it totalled almost 2,200 quintals! We were up with the top highliners! It was the largest trip I'd ever caught since first going as master. Only two schooners had more. We were in for six or seven days taking out that fish and enjoyed the time at home.

But I was still worrying a lot, and, before sailing on the next trip, I saw Doctor Burke in Grand Bank. He knew what every schooner had, but he didn't know how we got it or how down I was until I came up. Right away, he asked me, "Skipper, have you been worrying this spring?" "Yes, sir," I said, and I told him my troubles. "I know you're run down," he said, "I'll give you some medicine now and, before you come off the next trip, you won't feel like you're feeling now." I took his medicine, and a baiting of caplin and sailed. And we had the biggest trip, over 1,300 quintals!, on four caplin baitings. Before going home to discharge, I decided to fish a squid baiting. There was plenty of space in the hold, so one must make hay when the sun shines. With this big caplin trip, it seemed as if I was making progress. Bait was scarce that fall for our last trip on the Labrador, and some vessels got none at all, so our trip was a bit short and we arrived back in early October. But we ended the season's fishing with a large voyage, and did the same the next year. I was safe from then on in the *J.E. Conrad*. I was happy and so was my crew.

Chapter Six

Just a Six Hour Journey

A banks schooner had to be cleaned, repaired, and painted annually. This usually occurred in summer months between baitings or at voyage's end, before shifting it to either a winter mooring or to coasting service. To care for its hull, the schooner might be rolled on its side by shore lines attached to its mastheads. Drydocking or hauling out on a slipway were easier if available. This happened in November 1938 when Arch and a five men departed Grand Bank to sail the *J.E. Conrad* to a slipway at Burin.

They expected a six hour passage around the foot of the Burin Peninsula. Arch's recollection of this journey dispels illusions that sailing in Newfoundland's nearshore waters involved few risks and stresses. It also illustrates, as the famed seafarer and write Joseph Conrad — who is not the *J.E. Conrad*'s namesake — observed of similar experiences, "life itself...a symbol of existence...You fight, work, sweat, nearly kill yourself, sometimes do kill yourself, trying to accomplish something — and you can't. Not from any fault of yours." You can't move a 170 ton schooner to its nearby destination.

His story conveys the skipper's burdens in specially trying circumstances, how he attempts to manage them, the anxiety of loved ones ashore, and the experiences seafarers leave upon making port. When weary crewmen step ashore, their small clothes bundles in hand, the joy of turf underfoot and home help calm memories of recent hardships. Refortified, they are soon ready to sail again, expecting a better journey than the last.

1938

When I arrived home from our voyage in early November, 1938, Mr. Carr, the manager of Grand Bank Fisheries, asked me if I could arrange to take two of the men who were with me in the *James and Stanley* in the fall of 1936, when we had to jettison the deck cargo of oil in that storm northeast of the Funk Islands, and go up to St. John's to testify in a Supreme Court hearing over the incident. The owner of the oil we had to jettison to save our lives and the vessel, he claimed it wasn't right. Why should he have to be the loser, lose his summer's work? So he made a Supreme Court case of it.

We had planned to take the *J.E. Conrad* to Burin for docking before mooring her up for the winter, so I talked it over with her owners (J.B. Patten and Sons) and they agreed it could be done when I returned from St. John's. We went down to St. John's by steamer, and it was near the last week of November before the court proceedings[1] were completed. While I was in St. John's, a coasting crew of five men back in Grand Bank cleaned, painted, and fumigated the *Conrad*.

I arrived back in Grand Bank on the 24th of November, and, the following morning, Thursday, boarded the vessel to sail her to Burin, with a crew of six men.[2] If you have a half decent time at all, it's just six hours from Grand Bank to Burin in a sailing schooner. But there was barely a draft of wind, and it was calm all day. We only had the current that goes out through the bay to drive her, and it took the whole day to get from Grand Bank to the entrance to the bay, between Green Island and Dantzig Point.

It was just a little after dark by the time we got out to Green Island, when the wind suddenly sprang up from the

The Western Marine Railway dockyard, Arch's Burin destination in November 1938. c. 1929.

east. It got dirty pretty quick. I saw the lights up in St. Pierre and could have gone in and sheltered there, but we were bound for Burin and I wanted to get there and thought we might head off the wind. I had a feeling that there was a storm coming up because the barometer was going so high. But I thought we'd make Burin before daylight and get in down there.

We were in very close, hugging the shoreline because the wind was blowing off the land, easterly, and the sea was smooth. But it got so bad after midnight, we had to take in the mains'l and jib. She had no bowsprit, so all you had to do was just lower the jib down to beyond the barrack head.[3] When we had the mains'l in and tied up, and the jib and, later, the jumbo down, I ran off under the fores'l for about twenty miles to get some sea room. Then we just hove to with the fores'l up so she'd dodge in the storm. And we laid like that until daylight, when we put on the storm sail[4] and jumbo. Later that morning we added the jib.

The wind had changed from east to west by now and it

was still blowing a gale. But I told the boys that we might make headway and get down to Burin before dark. I figured the wind would be around from the west and fair before the day was gone. But the barometer kept falling, the wind went farther to the east, and we couldn't head outside the lee of the land. And there was a dense fog and heavy sea running all day.

If I could just see one rock I'd know where I was

It was blowing a gale, and we came in almost close enough that you could smell the land. But we couldn't see it. I knew it was there, and I figured, if I could just see one rock, I'd know right where I was. Then I could either go on around to Burin or get up to St. Pierre or somewhere. But now the second day was gone, and it was beginning to get dark, so I told them to take down the jumbo and jib, we'd run off again about twenty or twenty-five miles and heave to. Because I knew we were in on the rocks.

Just then I saw this sea break, right on her, almost under our bow. You couldn't see twice the length of the vessel for the wind and fog, it was so thick. But as quick as the reef broke, I knew it was a sunker, a breaker, up by St. Pierre, called the Lost Child, and I said, "Oh, boys, there's the Lost Child! We'll get up in St. Pierre now. Get the jib on as quick as you can!"

The wind was in our direction, and the barometer gave every indication that we were going to have one of the big gales, so we lost no time getting the jib on her. We'd be in St. Pierre in no time at all. We weren't going more than ten minutes when St. Pierre was straight ahead. Although we had a large vessel with a small lot of sail on, I knew we would get up in the harbour far enough to get out our two anchors and plenty of chain — if nothing gave out.

But now the wind was around to the southwest and blowing right smack out through St. Pierre harbour. We made a tack and steered over by St. Pierre on one side, let her come around and steered over on the other side — doing very

Just a Six Hour Journey

good — then another tack, and then the darn jib gave out. The jib halyard broke and down it came.

Now we only had up our storms'l, fores'l and jumbo, and, with a 170-ton schooner, you couldn't beat upwind. If the wind was a bit more abeam, and we could do a little bit of leaning, you could sail it through. But we had to tack straight into the wind. Well, it was dark by then, and the wind was up to about sixty-five or seventy miles per hour. And the barometer was down to twenty-eight.

It was like committing suicide whatever I did

I had about five minutes to make up my mind, whether to run in Fortune Bay or put her head out to sea. It was like committing suicide whatever I did, to run in Fortune Bay that night, when the nights are long, in a sailing schooner, in a storm of wind right in through the bay — there was no radar then. Why, you'd be up the bottom of the bay before daylight. That was on the 26th day of November. "My Lord." I said to myself, "What a place to be in, and on our wedding anniversary!" I made up my mind quick, to put her head out to sea.

She was hove to under the storms'l, fores'l, and the jumbo, with the wheel lashed, just jogging to windward of Green Island, close by all the reefs. To get down around St. Lawrence, we had to go right out around, about thirty miles, to clear Cape Chapereau (Chapeau Rouge). She was heading three points of the compass outside the stretch of land from Green Island to Cape Chapereau, and we were off about seven or eight miles. We were moving straight along the shore like that for about an hour, when the fores'l halyards parted and down it came. I went down in the focs'l with another man to splice the rope that parted. When we finished, the other five men got up in the rigging, one over the other, to pass the rope up. The top man got right up the masthead, and they had to shout to one another with every bit of breath they had.

When the first man was going up, he said, "Skipper, I can see a light there on our lee." He and I knew what light it was:

Lamaline light. The fog and rain had lifted a bit by then and you could see a little distance, so we knew where we were. It was about two hours before we got the fores'l back on her, then we had the riding sail, fores'l and jumbo again, and the wheel lashed.

I didn't let anyone else up on deck after we had the fores'l back on her. I told the men to go below because it was blowing so hard that she was washing everything right over the deck. The men were all in the cabin, nobody in the focs'l, and I stood in the gangway while she dodged on out.

We had a log out and hauled it in every hour. After the first hour it registered three miles. The next hour she logged another three miles, and the next another three, for nine miles. When I was splicing the rope down in the focs'l earlier, before we had the fores'l back on her, I thought to myself, "Anything can happen." We might have been in so close that she'd be up on the rocks at any time, because the shoals are so far from the land off Lamaline. But I didn't say anything to the men.

She had logged nine miles in three hours, so I figured she did another three miles during the fourth hour, about three o'clock in the morning. And we had at least another three or four hours to go at that speed before we'd be out of danger.

The big anchor we use when fishing in the summer was out on the bow, not even turned in over on the deck, because we were only going around to Burin and it came on us so quick. I went forward, found it was secure, not washed off, and rested my hand on the foremast and said to myself, "something has to give out. It can't stand it much longer."

It was blowing so hard it nearly took the breath right out of me. While I was standing there with my hand on the foremast, the fores'l, a brand new fores'l just last spring, made of what we call three-ought duck — canvas, the heaviest kind of canvas — it went just like matchpaper. Not a piece left as big as a tablecloth. Now all we had was that jumbo and the storms'l on a 170-ton schooner! Normally, she'd make a lee drift with only the storm sail and jumbo. But the wind was strong enough that she didn't, she urged along with the side laying on the land all the time. But I didn't know that then.

"Well," I said to myself, "we'll never, never do it. It's impossible." And I thought about everything that night. One thing in particular went through my mind: I have three children, and my wife was pregnant with the last one, our youngest. And I only had about one thousand dollars of insurance on my life.

I never got discouraged, especially with the men

By four o'clock in the morning I was getting about beat out. She was still in the water, but I don't know how. I told the mate to come up and stay in the gangway. That was all you could do, because the wheel was lashed and there was no use to get out on the deck. She was like a sunken rock. And I said, "She'll come out of it. Don't worry. She'll come out of it." I never got discouraged, especially with the men. They didn't know how serious it was. But they knew where we were because we were trying to get to St. Pierre, and they could see the light.

There was an ice house that stood in Grand Bank for forty years, and it blew down that night. And there were roofs blown off of the houses. (Mrs. Thornhill never forgot the squalls at night: "There were an awful lot of storms then, and Grand Bank was right open to the sea and the wind would come. The old house was perched high, and the wind was high, and, oh, did it ever crack and creak and make such a noise. It sounded worse because of him being out in it, and I often laid awake. But you couldn't let the children know how uneasy you were. I kept a lot inside.")

I went and laid down on the cabin floor. They all had on their oil clothes, and I laid down with the rest of the boys with mine on. It was between about four and four-thirty in the morning, and I was only out on the cabin floor for about twenty minutes, when the mate came down and said, "Skipper, the wind have chopped right in the opposite direction, and you wouldn't be able to hold a sharp edge knife to windward. There's a living hurricane now, right from the narth." A northeast wind was blowing right smack off the land. I said, "Thank God." I never looked up. I knew that if she never struck the rocks to that point, she wasn't going to strike then.

My Lord, she was everywhere except on the rocks!

The foregaff was up and the foreboom all up. The duck was gone, but that was still up, swinging back and forth. I said, "Let everything crack. Let it stay where it is." And I went and took off my oil clothes. Everything was wet. Every mug we had on board, we usually have hung up on the beams, was full of water or broken. Every dish in the cupboard was full. The bunks were full of water as well. Oh, my Lord, she was everywhere except on the rocks. I said to the mate, "Now, settle your watch. Take a man, and carry on your watch." We had four-hour watches then, when you were coasting. No worries now, we are driving right off over the Grand Banks.

At nine o'clock that morning we got up and had breakfast and I said, "Now, call 'em all out. Call the men out." And we took down all the wreckage, the booms, and snugged it away. We drove about one hundred miles off over the Grand Banks on Friday, our wedding anniversary. I'll never forget it. And by Saturday evening the wind started to die down a little bit, and on Sunday morning it was around to the east. There was still a big swell in the water. Now we got up, put the big mains'l on her again, and took the storm sail and put it up for a fores'l, and put the jib and jumbo on her and cracked it to her right for Burin. It was a bad looking sky, every indication of another easterly gale.

I said, "We'll make that tonight. We'll be in there before dark." But a thick snow came down in the last couple of hours as we got in close to land. You couldn't see a hand before you. But I kept her going, kept her going, and thought to myself, "we might see it." Soon the wind grew so strong we had to take in the mains'l and face it. Then it was over to, under the substitute fores'l, and jumbo again, and put her head up to the south. And there we were down near Burin now, in the same state we were in off of Lamaline.

About two o'clock that morning the watch came down and said to me, "Skipper, we can see a light." It wasn't very clear because there was a storm of wind, and I said to myself, "Yes, boy. That's a light in on the land." I didn't say it to him. I got up and put my head in the binnacle, in the light, and I kept my eyes on that light for about a full hour with a course towards it. I could tell he was moving a bit fast, fast, to the east, and I said, "O.K. boys, that's a big steamer. We're not quite so far in towards the land as I thought we might be," which was a good thing. That was a relief.

In my imagination . . . I saw the Cays fifty times that day

About an hour or so after that the wind was around, right smack into the southwest, right in Placentia Bay, and the barometer dropped below twenty-nine. I knew by now that we were in for another bad one. The next day, there we were in Placentia Bay. Now, Placentia Bay is a bad place too. We had to head toward Cape St. Mary's with only the little storm sail and jumbo to get us out of it. Going across from Burin to Cape St. Mary's, there is a place off of St. Mary's called the (St. Mary's) Cays — a bad place, about six miles off. Breaking all of the time over the worst kind of breakers. I suppose, in my imagination, that I saw the Cays fifty times that day. I figured that the wind came from the north-northwest as fierce as it did the other night when it blew us out of danger. We weathered down, and drove another twenty-four hours off over the Grand Banks again. The next day, around two o'clock in the morning, the wind died out, and we got up and put that big mains'l on. But just as we had it on her she gave a surge in a swell and the jiver — the blocks and everything where the sheet goes in — burst. And the main boom went right out across the rigging. It's a wonder it didn't clean her down. By that time, it was blowing a gale and we were hove to again. By two o'clock that day we got that big mains'l tied up and snugged away.

By now we're all getting broke up. The men had been working hard and in wet clothes nearly ever since we left home. Wednesday afternoon the wind moderated and we made some progress toward land, but we're still off quite a distance.

This was our seventh day away from home now, and it was only supposed to take us six or seven hours to make Burin. After we were out five or six days, my wife called Howard Patten, the manager of the firm, J.B. Patten and Sons, up to our house. She said, "Mr. Patten, don't you think that we should do something! Don't you think that you should do something? Arch is not lost. They are not lost. He's out there with his sails gone. If they had been lost on this shore someone would have seen some wreckage. He's not lost," she said, "and I think we should get in touch with the Department of Natural Resources (in St. John's) and get them to get steamers crossing the Atlantic on the lookout."

"Oh," he said, "Mrs. Thornhill, I don't think there is any need of it. I don't think there is any need for alarm. Arch got good experience." "Yes," she said, "he could have good experience, but what's the good of experience if he has no sails? Mr. Patten, what would you do now, if your brother," who was skipper of a schooner for a few trips that year because the regular skipper got sick, "was out in that schooner instead of my husband?" And he went right to the door and said, "I don't think there is any need of it, Mrs. Thornhill."

But she sat down, wrote out a ninety-word telegram, and wired it right down to the Department of Natural Resources, and within fifteen minutes she had an answer back from St. John's: "Will alert all the ships in the Atlantic and also send out one of the coastal boats." About half an hour or so after that, Mr. Patten came up to the house again and he said, "Mrs. Thornhill, I think we should send a telegram to St. John's and tell them about the Conrad and her crew." She said, "You don't have to do that now, Mr. Patten," and she showed him the telegram.

A good many people . . . thought the *Conrad* was gone

(Mrs. Thornhill recalled that night: "I thought it was time to worry. Or time for something to be done. So I went and sent the telegram myself, because I felt nobody was doing anything. Captain Alexander, an old skipper, was sitting in the Post Office when I sent the telegram. Later he said I looked a bit worried. But I think all, or a good many people in Grand Bank, thought the *Conrad* was gone. Because there were so many tragedies then in Grand Bank.")

My mother happened to be out there at the time just for a visit, and I had my brother, Wils', aboard with me. Well, she had us lost. The coastal boat was in Grand Bank that morning, and mother went home to Anderson's Cove on the steamer because she knew my brother and I were lost.

They put the alarm out and this big liner was heading for England. It's almost too much to believe, but he changed course, came in over the Grand Banks, and just at dark he sighted this schooner, water-logged, decks awash. All of her dories were broken up in the storms. It was too dark to see her name, but they went alongside and put out their lifeboat and took the crew off. When they came aboard the liner, her captain said to the schooner captain, "*J.E. Conrad*, hey?" But he said, "No, sir, this is not the *J.E. Conrad*." "Not the *J.E. Conrad*? Well, an SOS call came out, flashing around about the *J.E. Conrad* was overdue from Grand Bank to Burin." The schooner captain said, "This is the *Allan F. Rose*, this schooner. We left St. John's about seven or eight days ago." He drove off in the same storms I was off in and his vessel sprang a leak. The steamer that came to look for us ran into her and saved the six men aboard her. They weren't more than off before she went right down, sank right down.[5]

The sea was too rough to put our mains'l on during the night, but we made fair progress using our storms'l for a fores'l and our jumbo and jib. We were getting in towards land again when we sighted a schooner, and I said, "Oh, boys, there's a schooner. Let's put our flag bottom-up. If anyone is in distress, I'm in distress because I haven't had a bowel movement for eight days." When we fumigate the schooners in Grand Bank, they take out the medicine chest and everything else. But we forgot it and had nothing aboard, all the medicine was ashore when we sailed. And I said, "Put the flag bottom-up because, if ever anyone is in distress, I am."

He came down alongside and hove to, and he put out the dory and my brother and I went aboard. Her skipper was Murley Hollett, from Burin, and the first thing I said was, "Murley, boy, before I tell my story," and I told him about my distress. I said, "Have you got any medicine here?" "Oh," he said, "I got a bottle of castor oil here in the medicine chest." I took the bottle and put it right down. "Now," I said, "I'll start talking," and I told him the story. "Well," he said, "you shouldn't be very far from St. Pierre because we left (North) Sydney yesterday morning." He was going to Burin with a load of coal.

So I went back aboard the *Conrad*, and while we were hoisting up the mains'l, my brother looked and he said, "Oh, skipper, there's land there. Look." It was St. Pierre. But just as we had it hoisted and were making the halyard fast, the block eighty feet aloft broke, and down it came with the mains'l. I'll never know how near we were to having someone killed. We didn't take long getting things repaired and the mains'l on again. "Well," I said, "we've had enough of this. Let's get it on just as quick as we can and head for Grand Bank now." I knew they had us given up for lost; only a six-hour run.

We weren't inside Green Island light before the wind came out from the northeast, and this awful medicine was working on me, and there came the snow. It was early December now. And there I was on the draw bucket[6] every now and again — there were no toilets aboard the schooners then.

"Well," I said, "one thing is sure. The two masts will go out of her this time before I run out through this bay again. I'll keep it on her until she blows level with the deck." And, boy, she breezed up, and she breezed up, but when we weathered Grand Bank, it was blowing.

When we were fair for the harbour, we lowered everything down and ran right in through the gut under bare poles. And we went right up in the bottom of the gut and I said, "Now, boys, don't tie up anything. Don't snug anything away. Just hitch her on with the line. Everybody go to your homes."

An awful lot of people never came home

I went up to my house with my brother, and there they were: my wife and three children in one bed. I think she was the only one in Grand Bank who never gave up on us. The rest had blinds and everything down for us.

When I went in the room that morning and the three little children and the wife were there in bed, 'twas a proud thing to see. And 'twas proud for them to see me too. In those days, the little ones, when they go to school they're always looking for the day mom gets a telegram from dad. She pinned it up in the window. They came looking for it, day after day. I've often thought about it now. An awful lot of people never came home. A lot of people got lost. How sad it must be.[7]

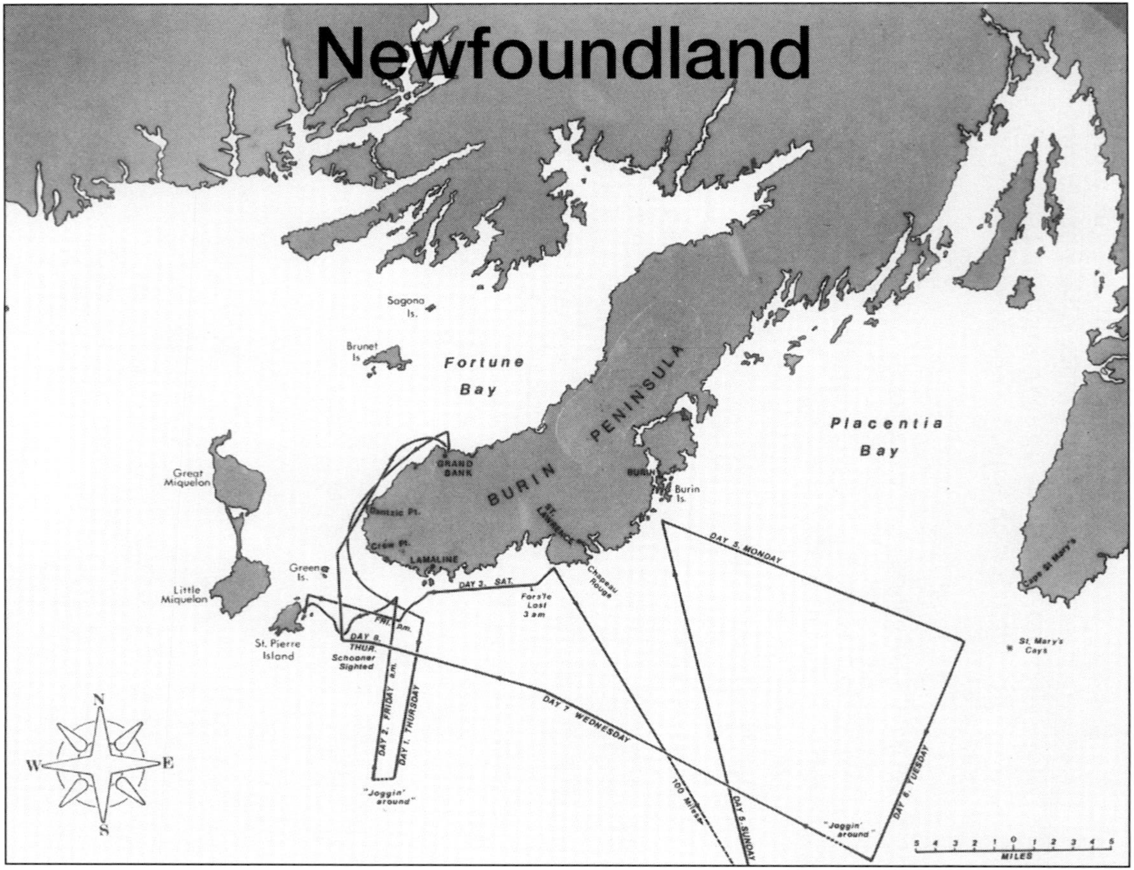
Newfoundland
Sagona Is.
Brunet Is.
Fortune Bay
BURIN PENINSULA
Placentia Bay
GRAND BANK
Burin Is.
Great Miquelon
Little Miquelon
Green Is.
St. Pierre Island
LAMALINE
Chapeau Rouge
DAY 3. SAT.
DAY 5. MONDAY
DAY 7 WEDNESDAY
DAY 6. TUESDAY
DAY 5. SUNDAY
DAY 2. FRIDAY a.m.
DAY 1. THURSDAY
DAY 8. THUR. Schooner Sighted
"Joggin' around"
100 Miles
St. Mary's Cays
Cape St. Mary's
MILES
N
S
E
W

Chapter Seven

Never Lost a Line, Never Lost a Hook

In this chapter Skipper Arch accepts command of the *Florence*, an American-built banks schooner. Most of his crew from the *J.E. Conrad* followed him. Well they might, as the *Florence* was scheduled to have an engine installed. Her captain's characterisation of her as a 'dummy' schooner may ring oddly. Skipper Arch reminds us, however, that its owners and crew recognized the advantages of power for its operations and integration with 'making fish' onshore. It promised more profit and earnings.

Banks fishermen knew and compared skippers by reputation for landings and earnings, vessels commanded, seamanship and crew treatment, and men lost. Comparisons were sometimes discordant. For example, south coast banks fishers often remarked that the best 'fish dog' or 'highliner' banks skipper known to them had also "lost a banker's crew". However, if banks fishing's tests and risks were attractions to some men, its hardships soon stripped them of illusions. They knew lives could be and were lost in countless uncontrollable ways. Given the opportunity, they would have sailed in dory with the highliner skipper themselves.

Despite competition for fish, vessels, and crews, banks skippers were family men rooted in coherent, small and egalitarian communities of hard working people that depended greatly upon their fishing enterprise. Relations between skippers from the same companies, fishing centres, communities, and coastal 'neighbourhoods' were often close and even familial. Skippers and crews knew their landings meant work for hard pressed people ashore who made the fish for them. Hence, if opportunity permitted, and it did not harm a skipper's own catch, he might share information on fish locations, as we see here.

It may surprise some readers to learn that their hazardous and bone tiring competitive work regime remained integrated with landward religious practices. Most outport people of Arch's generation treated Sunday as the Sabbath, as a day of prayer and rest. They did the same on the fishing grounds. Hooks and trawl lines were put aside, men slept in, and later gathered in or around the cabin and foc'sle for prayers and singing. As Arch says, "they were good religious people then." It made sense. There was an immediacy to forces beyond human control. Men thought more about them then.

1939-1940

In 1939 we had our third good voyage in this vessel, with smooth sailing all the year through. Our second son was born to us during the year, and our family now had two girls and two boys. Everyone was saddened by the declaration of another world war, and things looked serious. After our fishing voyage was finished we took on a crew of six men and went freighting until after Christmas.

We arrived at Grand Bank and tied up the vessel for the remainder of the winter. During the winter months our owners decided the *J.E. Conrad* would go into the foreign trade this season. I was very sorry, because I was proud of this fine, large, bank fishing vessel, but there is always a lot to be thankful for. We had three excellent voyages and I managed to pay off my loan in these years.

The last ten years had been very bad for bank fishing and most other businesses. Some fishing companies had gone entirely out of business in Grand Bank, and in other places in Fortune Bay. Our owners asked me to go captain of the *Florence*, another one of their bank fishing vessels, and I accepted.

During the last two or three years, many Fortune Bay banking vessels had acquired small, light-duty engines, which was a big improvement.[1] Our owners advised me that, if all went well, they would have one put in the *Florence* after our voyage. She was nearly the last sailing, bank fishing vessel in Newfoundland. There were trying times for me during the past year, especially when laying becalmed and another vessel just steamed past. But we had smooth sailing and did well, so I guess someone has to be last in every game.

The *Florence* was a nice vessel, and had given a good account of herself over the years. She was built in the U.S.A. and purchased in Belleoram by Harvey and Company. All vessels built in the states then were low set and took a lot of water on deck, especially when getting well down with fish. It was quite a change for me, as it was the first American-built vessel I had sailed. It wasn't long before a nickname was put on sailing vessels after the engines came along; they were called the "dummy" vessels. So I had to try my luck for at least another year in the old dummy. I had nearly all the same men who were with me in the *J.E. Conrad*, which meant a good crew. On March the 10th, 1940, I left Grand Bank, sailed in Fortune Bay, and came back with my crew to finish our preparations. By March the 20th, our supplies and bait were on board, and we sailed for the banks. We returned from the banks after about three weeks, and discharged our catch at Grand Bank. It was a fair trip, but the vessels with power did better.

When the fish were discharged and fresh supplies on board, we sailed for Fortune Bay and took on another supply of bait. It was very calm when we came out through the bay, and at least half a dozen vessels graciously steamed past us, the poor, old dummy. We were quite a while getting to the Grand Banks, where we fished two more baitings. Our spring trip finished when we arrived at Grand Bank on June 1st with an above average trip. I was very pleased indeed.

Since the vessels have had power, they always fish their first baiting of caplin and come to their home port and discharge, then continue on and finish their trip, usually with two more caplin baitings. This was because July was always a bad month for drying fish, and many, many caplin trips have spoiled on the flakes and beaches owing to rain and fog at the time. For many, many days in some years the fog was dense, and it rained continuously. "Swifton's days,"[2] it's called. But in the old folklore it was always "dog days." Before vessels had engine power, there was no other choice, only fish your trip, come home and take a chance on getting your fish well made.

Before we left on our caplin trip, being the last vessel, our owners said they wanted me to fish one baiting, come home, and wash our fish, if we could get 500 quintals. Well, that was always considered a good first caplin baiting. Nevertheless, we did our best. The *Florence* was now always the last vessel getting to her destination and to home port this year, owing to being the only dummy around. But we got the 500 quintals, and 100 more, and set sail for Grand Bank, some 250 miles from our fishing position on the Grand Banks. Despite the distance, we hit a good time and swung the four lowers all the time, until we arrived. No time was lost discharging our fish.

When we were going into Grand Bank, we met the last engine-powered vessel to discharge her catch coming out. Others had landed and departed days before. Not many had over 600 quintals, and some never had quite as much, so I was pretty pleased at what we had achieved so far in the last, old dummy.

A couple of days later, we were on the way again. A baiting of caplin was quickly secured, almost in Grand Bank harbour, and two more caplin baitings were fished, but we found fish scarce. Only about 600 quintals were caught on the two baitings, but it was still considered a big caplin trip. With the 600 landed on our first baiting, we had 1,200 quintals. Prospects looked good for getting power in the old dummy after the voyage.

With only 600 quintals on board, I decided to go to St. John's for some frozen squid bait and proceed to the Grand Banks again. It was now about the tenth of August month. There was sufficient time to fish another baiting, and our frozen squid bait kept longer than the little iced caplin bait. On Saturday morning, we were within ten miles of our destination, it was blowing quite a breeze, and we had already shortened sail a few hours before, when we saw a vessel coming fair towards us. Getting so close to our destination, we would normally heave to and board the coming vessel to get all the fishing news. But the wind was too strong to put a dory over the side, so we kept our two vessels as straight for each other as we could. Both were clipping along pretty good. In fact the other vessel had her sails on and engine running at the same time. It was Captain Alex Smith and the *Nina W. Corkum* from Grand Bank.[3] I didn't have time to ask him for the information I wanted, but Captain Smith knew what I was going to ask him. He hollered from the top of his voice with his speaking trumpet, and said, "Go on the same direction for eight more miles and then anchor."

We could see his vessel was loaded and carrying home fish on deck. We went exactly eight miles in the same direction and anchored. The next day, being Sunday, we did not fish. It wasn't long after midnight Sunday before the men were all on deck around the cabin, baiting their gear. At daylight, it was a lovely moderate morning, and all gear was set. The men came on board the mothership and got their snacks and their bait, to go out and underrun their gear. In no time at all I saw the ten dories coming, all loaded. We're not counting fish these days.

My cousin and I were up to our vows and never counted

fish after we went captain, and by 1940, there's not a vessel counting fish. Around the last part of the week, my cousin, Captain William Thornhill, came along in the schooner *Eva A. Culp*. He and my older brother, who was a member of his crew, came on board to get the news from us. We had been fishing five days now, averaging over 100 quintals a day, which is excellent fishing. After they got all the news from us, and we got the news from home — they had come directly from Grand Bank — they went a berth from us and anchored.

We were on a lot of fish, and just about fully loaded in that one place without weighing anchor. The fishing was still good on Saturday, which gave us six full days of fishing there. Unlike Nova Scotia vessels, which sat somewhat high out of the water, when just about loaded, our American vessel looked pretty low in the water.[4] We nearly filled the hold and took out the small amount of salt we had left and put it in the cabin to make room for another day's fishing on Monday.

We were good, religious people then, didn't fish on Sundays. . . .

On Saturday, the last haul of the day, I said to the men, "We'll take in our gear now and shack it up." It had been out all week, and many of the sudlines and hooks were torn off. We'd take it in on Saturday evening, after our fish was salted and put below, and repair it over the weekend.

We were good, religious people then, didn't fish on Sundays, and were glad when that day came. Then we'd just lay around. As a matter of fact, we'd be tired all during the week. On Sunday the cook would call out, "Breakfast!" about an hour after daylight. After breakfast we'd turn in again, sleep 'til about dinner time, when the cook sang out again. Everybody had plenty of sleep by that time, and we'd just chew the fat, sit around, tell yarns and one thing and another, until supper time. Almost always, after supper, after the cook got settled away and the supper dishes cleared up, everybody'd stroll back, and get around the cabin, and sometimes the focs'l, and sing hymns. Somebody would probably have an accordion, and he'd play. Twenty-odd men would be back in the cabin, and, oh, we'd have a real sing-song.

Everyone was in good shape again by Monday morning. About one o'clock in the morning, we got up at the usual time to bait our gear. After an hour or so the cook sang out, "Breakfast!" Just before daylight everyone was finished, and we probably waited half an hour or so for daylight before we put our dories out. Every dory went out in the same direction they were in during the past week, and set their gear, about twenty-six lines a dory. And, when they got their gear set out, they all came aboard for lunch and to cut up their squid for bait, to go out and underrun the gear.

'Twas one of the best mornings that ever shone out of the heavens, and the Atlantic was as smooth as a pond ever was. Just as they were about to get in their dories to go to their trawls, I went in the cabin as I always did and glanced at my barometer. Although it was the finest kind of a morning, I saw the barometer had gone down a tenth within minutes. We still had no radio then, so we couldn't get a forecast. I said, "Hang on for a few minutes." They looked at me, and at each other, and wondered why. I kept them there for ten or fifteen minutes. And I went in the cabin and tapped the barometer again, and it jumped another tenth. That was almost two-tenths in the last fifteen minutes. So I come up and said, "We'll hang on a while, anyway."

The old man must of had a bad dream last night

I knew the crew was wondering why, especially on this beautiful morning. Within half an hour I could see a few clouds in the sky, and the barometer was dropping faster all the time. But no wind. Not a draft of wind. I said to the men, "Hoist in the dories." They looked at me, wondering. I heard one man say, "The old man must have had a bad dream last night." They hauled up and hoisted in all the dories. About fifteen or twenty minutes after that you could see a puff of wind coming on the water, and, within an hour, the sky was beginning to look bad. You couldn't see the sun, it was all clouded in. Then I gave them orders to grout (grip) down the dories. We had gripes to put right over the dories and gripe them down on the deck, as we always did if there was a storm coming. All the crew knew then something was brewing.

An hour after that, I told them to batten the hatches. The wind started to come now, started to breeze up fast, and they didn't wonder anymore. They knew I used to watch the barometer closely. With about 1,300 and 1,400 quintals of fish in her, she couldn't take much more. Another fifty or sixty quintals and she would have been log-loaded.

By twelve o'clock the wind was beginning to howl at gale force. And in about two or three hours, it was blowing so hard, she went adrift, and started to drive broadside, right down. Shortly after midday, we slacked out two or three lots of cable — every bit we had, about 250 fathoms. And the water was only about 37 fathoms deep. The riding sail we use on the mainmast when anchored was already torn to threads, so she was driving back, driving back.

Within three hours after I told them to hang on, not to go out to their dories, it was up to about sixty or seventy miles per hour. And there were some big seas. You wouldn't think it could come so quick. And, now, with the schooner so deep — we had practically a full load of fish on board and some of our salt in the cabins — she's driving broadside to it, she's washing pretty good. Three men and I were on deck, to our waists in water all day. The rest all sat up on the main boom, above the cabin, and up in the rigging.

A fishing vessel carried six punchions on deck to put cod liver in, where it rots to oil. The six butts were full of oil and liver, so I told them, "Get a hatchet, pin mauls or something to beat the heads out of them, for all the liver and oil to come out to lighten up the vessel, and it will smoothen the water at the same time." Then I told them to beat the bulwarks off on the leeside, to beat everything right off so the water would get through freely as it came over.

A dory wouldn't have lived two minutes that day

About twenty years later I had a cook with me in the *Makkovik*, one of the small government boats I was in after I gave up fishing. He was one of the men in dory with me on the *Florence* then. We were talking it over, about the past, and he said, "Skipper, I'll never forget what you said that day." And I said, "What was that, now, cook?" He said, "You looked up to me and said, 'George, I wouldn't give two cents for our lives'." And I said, "Yes, boy, I guess I said it. But, as a rule, I wouldn't say it for anything." I said it only to him. I didn't know I did, but I must have, if he said so. Because, if one of the hatches would have moved, washed off two inches, the slightest bit, she'd have gone down just like lead with all of us. We wouldn't have gotten a dory or anything off the deck. Anyway, a dory wouldn't have lived two minutes that day. That was my biggest worry. Not about the vessel weathering the hurricane, but if the hatches would keep in place.

You could barely hear a thing because of the howl of the wind. It was up to hurricane force. I was walking forward, and, right in the midst of it, I heard a screech, a yell. I looked around, and there was a man out in the water. A sea, one of them monsters, came down and struck, and washed him right out overboard. He was about 250 pounds, and three men had grabbed him. The schooner wasn't going ahead, she was driving back, making a lee drift, and he came back the same way. If there had been any sail on the vessel, she'd have been racing ahead, and he would have drowned.

The three men had a hold on him, but they couldn't get him in over. I was scared someone else would go while trying to save him, so I got back with them, and with my little bit of extra strength — I didn't have much, but with the little bit of extra pressure that I put on we yanked him right in over the rail.

We towed him along to the cabin gangway. It was all closed up because so much water was coming right in over. We watched our chance to pull the gangway open, and helped him in the cabin. I said, "Go down, uncle Bill (his name was William Riggs). Go down, boy, and take off your clothes. Change your clothes, and say your prayers." He never got on deck again that day.

That wind lasted eight hours, driving down, up through the northwest. It was southeast, direct, and she was driving. Then it chopped from the southeast to the northwest just as quick as I'm saying this, and we were driving back direct to the southeast. And the barometer is on twenty-eight now. I can't say I ever saw the seas as big in all my life as on that day. The whole Atlantic that we could see was like reefs and breakers around the land, going up so many feet in the air.

We drove to the southeast for about eight hours, so the hurricane lasted sixteen hours. 'Twas in the summer time, and the days were long, about sixteen hours of daylight. It wasn't dark until after ten o'clock. But it was getting late now, and the wind started to moderate. The barometer began to come back about half-an-hour or three-quarters of an hour before the sixteen hours ended. When the barometer comes back, that's when you'll have the hardest wind.

Shortly after dark it began to moderate a bit quick, and the barometer was back as far as it went down, and by about midnight, the wind was below gale force and the sea had become quite smooth. So I said to the men, "We'll try to get in our anchor now," because one anchor and all that cable was still out. We had towed it all day. We heaved in the cable, got the anchor up, and found its wooden stock had broken off in the storm. That's why she went adrift. It was a good thing, otherwise she would have sunk at anchor. Other vessels have done so. Of course, we would have tried to cut the cable before she did. I had done so once before in another vessel. Anyway, a couple of hours later the wind was about 40 miles an hour, and it was moderate enough that we were able to put in another stock, and we had the anchor fetched up on the bottom again.

We had eleven dories' gear out, twenty-six lines a dory

By midnight, there was just an ordinary breeze, but a big sea, a big swell. The next morning, just at daylight, we got up, put all the sails on her, and I said, "Now, we're going back to get that gear." We had eleven dories' gear out, twenty-six lines a dory. I didn't say, "We're going back to look for it." "We're going back to find it." I said it as more of a joke, and the crew said, "Skipper, do you think we will find it?" I said, "We may. Never can tell."

We put the sail on and started beating to windward toward where I figured the gear was. I knew we couldn't be much out because we drove up to the northwest for eight hours and then we drove back to the southeast again. I used my judgement on how far it was from us, not more than five or six hours just going to windward.

There were six, seven, or eight men in the rigging looking for the little buoys, the black balls with little flags on them, we had on our gear. By and by one man hollered out, "Skipper, I can see some over here." So we got right up then, and anchored about where she was when she went adrift. The men went on out and took the gear in. Never lost a line, never lost a hook. We got the eleven dories' gear and all the buoys.

We only had enough bait for about one more day's fishing, but we had all the bulwarks beat out of her, so I said, "We'll call it off now and go home." We had just about a load then anyway, 1,400 quintals altogether. We had secured 800 quintals in six days and never weighed anchor, the old dummy. We came home then a couple of days after that, washed our fish out, and put it on the beaches.

Chapter Eight

Hang On a Little While

In this chapter Skipper Arch describes how dory crews drew for berths, i.e. the compass course on which trawls were set from the mother vessel when fishing at anchor. The description is "classical." The practice was used by banks schooner fishers from New England to Newfoundland. And it was used under the onerous count remuneration system Arch reports he abandoned years before. Yet the draw continued — for equity's sake. But in what sense? The count's excesses were replaced by a system that gave all hands an equal crew share in their vessel's catch value. Thus all hands might better "pull together". However, this change did not overcome work conditions created by wind and tide when setting and retrieving gear and fish. Who should have to haul back against the elements? The draw randomly distributed the advantages, difficulties, and hazards of wind and tide. No one was favoured when going from and to the anchored mothership.

A banks schooner flying under sail is a lovely picture. It is also the fisher's home at sea and a machine, a tool, for catching fish. If we admit it is a material and symbolic thing, we may hardly recognize or understand its many embodied meanings. Consider, for example, how their names — like *Bessie Macdonald* and *Florence* — anchor steps in Skipper Arch's career and evoke positive associations. It is perhaps obvious that names individualize and distinguish vessels from each other. But they do more. A typical list of vessels operating in Newfoundland waters earlier in this century reveals about seventy per cent carried personal names. Two thirds of these were after wives or children, and the rest were after males. The other thirty per cent had "tribal-regional" (e.g. *Alsation, Garnish Queen*), historical-heroic (e.g. *Dauntless, Admiral Dewey*), natural (e.g. *Coral Spray, Gladiola*), animal (e.g. *Leopard*) and other associations.

Does a preference for female names somehow explain or reflect the common practice of referring to them in feminine terms (e.g. "she's a fine vessel") . . . no matter the gender or other name? If so, what is this connection about? Interpretations might range from the ordinary to the exotic. For example, we might say that owners chose personal names — perhaps a wife or a daughter — that mark and perhaps honour the achievement of personal ownership itself. Somewhat like a medieval champion, the owner's skipper celebrates and fights for his master's house by sailing his namesake.

New schooners have long been launched and named simultaneously and ceremoniously with a "christening" and other ritual practices. The intent is both to invoke benevolent sacred forces and to inspirit the inanimate machine with the namesake's imagined positive qualities — including luck itself. The process is magical and spiritual. Most fishers saw banks fishing as subject to unseen forces. For example, in this chapter Arch notes how some of his crew worry about the dire implications of a rat seen swimming ashore as they sailed from port. The skipper might be mindful to insure the schooner was never turned against the rising sun. But what to do about the forces signaled by that rat?! These concerns and practices are also "classical" in being common among North American and European sailors and fishermen alike.

Vessel names and other do's and don'ts are efforts to influence otherwise indifferent fortune to deliver, like reproduction itself, bountiful catches and safe sailing. At the least, a name and its positive associations may subtly reinforce an owner's confidence in outcomes. What does the name mean to the skipper? If he hasn't chosen it, perhaps little at the outset.

In any event, when Captain Thornhill watches the *Florence* sink beneath the waves, her passing affects him deeply. She was the means to a living, his home away from home, his achievement, scarce capital equipment in hard economic times, a daughter's namesake, and more.

1940

When our bulwarks were repaired, I took her up on the Labrador with a baiting of frozen squid. Most schooners fished on a big patch of ground down there called the Round Hill (around an island by the same name), about three or four miles off Salmon Bight, where we harboured. It has about ten or twelve miles of ground, and was better than any other place off the Labrador for fish in the fall. The *Florence* was almost the only dummy schooner around then. The vessels with power could steam in and harbour in Salmon Bight, while we were always the last schooner to come in. I couldn't hang on in the harbour the way they did or they'd head me off and have all the good ground, so we anchored outside, and left again at about twelve o'clock at night to have our gear out on the good grounds before they did.

There was plenty of cutting lines from one schooner to another

But this one fellow, his crew cut us up, bit by bit, morning after morning, day after day.[1] We had to get out of it and get new lines from someone else. And that was only us. This went on for a long time. With a small patch of ground and a lot of schooners, whoever got their dories there first, they'd get their gear all cut, cut, as the others went along. They've been down there and wiped one another out. You would try to be the first to get there, otherwise, cut and get the gear and fish. It was a hard racket. But we weren't counting fish then.

There was plenty of cutting lines from one schooner to another. Not so much on the Western Shore, where you fish in about 100 fathoms, but a lot down on the Labrador where you have shallow water, from seven to ten fathoms. And sometimes it was your own fault. You would set your gear a little too close to another schooner. Perhaps you knew the other fellow was trying to hit it big where he took a lot of fish the day before, and you squeezed up as close to him as possible, then two or three of your dories got tangled up with some of his. If you weren't there first, someone would cut you up, because you had no business going there.

Back in Grand Bank again, we discharged our fish and took on food and salt, and went on out for another trip. As a rule, this would be our last trip and we would go up to the Labrador. But it was now about the 10th of September, and a bit late for that. So we went up the bottom of Trinity Bay, took some frozen squid from the bait depot ship, *Mallakoff*, stationed there and went off on the Grand Banks. There were few schooners off on the Grand Banks at this time. Most were on the Labrador and up on the Western Banks.

We went down and anchored at about the same place where we filled the hold on that last trip in early August, and there was a lot of fish there again. When we put our dories out, I said, "If we could get three or four days of good fishing weather now, get up to 400 quintals of fish, we'll have a savin' baitin'," because it was only going to be a short trip. And, boy, on that first day, we got 100 quintals, and nearly another 100 on the next day. Then there were two or three blowy days like you get in September, and a big tide, a big current running. Anyway, we fished a few more days there, made up the four or five hundred quintals, and had a good baiting. And I was very pleased with our voyage by then.

It was nearly time to leave for home, but we had enough bait left for two or three more days. The weather was coming on strong, so I decided to head towards home and stop on St.

Pierre Bank, about 100 or 120 miles from Grand Bank, and fish for a couple of days. Then we'd leave. That night, we struck a breeze going up and had to shorten sail, take in the mains'l and jib, and run her right before the wind. Sometime the next evening we were up on the St. Pierre Banks, and we anchored.

The next day was a gloomy, poor day, and we didn't fish. But the following day gave every indication of being good, so I called them out at about one o'clock in the morning, and they baited up their gear. We were going to fish under sail.

When we fished at anchor, on a spot on the Grand Banks, the dories set around the mothership. We'd make up so many compass courses, so many points apart, depending upon how many dories we have. Then each dory skipper would draw a ticket with a course on it from a sou'wester'. And then you'd go on your course, north, northeast, south, and so on, around the vessel. You'd moor the gear with anchors, take your bait, and go out and underrun it; when you hauled towards the schooner, you'd pass your gear over the dory, and it would be out fishing while you were aboard. Sometimes we would fish seven or eight days like that in one place — if you were lucky and struck the fish.

But today we fished under sail. We dropped our ten dories with their gear (every 200 to 400 yards) from the side of the vessel. Every dory set down the one course the skipper told them to sail before they left. Say, "Set southwest, now, today, boys. Everybody set southwest."

You would tow them along in this direction, each dory would go down, and they'd drop their buoy overboard and tow out their buoy lines. When you got a little distance from one dory, you'd drop the next one and go on until all your dories were out. And when the last dory was dropped, you would sail down the length, the reach of that gear, to pick them up again. After setting their gear, I'd pick up the first dory I dropped, and go back until I had picked them all up, in rotation, as I dropped them. I'd sail back on the weather edge, and by then the men would be finished with their dinner and you'd drop your dories again. Then they'd haul their gear in the same direction that they set it. That's fishing under sail.

After we dropped our ten dories on their gear, I began sailing back and forth through them, anxious to see if they were getting any fish. 'Twas a very good day, about fifteen miles an hour of wind, and I could see two or three other schooners fishing in the vicinity. They sat four tubs to a dory; we had about ten lines in a tub, about forty lines. When they had their gear set out, I took them aboard the vessel again, and they had their dinner while I beat up on the windward ends. By the time I got up there, they had finished dinner, and I dropped my dories again and they went on to take in their gear.

Everybody was feeling pretty happy when they left the vessel. We had a good voyage landed and we had approximately 500 quintals on board. My owners told me before we left home that as soon as our voyage was over we would go to St. John's, to have an engine installed. And I had already chosen five men, besides myself, to be the crew for the trip.

We were approximately 120 miles from home and there was every chance we'd be there around two or three o'clock the next morning. I had already told my men that this was going to be the last day of the voyage, the last day.

The men saw this rat come out from the cable....

I never was superstitious, but a strange thing happened when we were leaving Grand Bank on this last trip, and went down to Trinity Bay to take on bait. Just as we got out through the harbour and about 200 or 300 yards from the pier where we were tied up, the men saw this rat come out from the cable, forward. They thought they were going to have fun killing it, but the rat was too smart for them and jumped overboard and started swimming ashore. They threw things at it, but missed every time. I don't think any seaman likes to see a rat leaving the ship when leaving port. The saying is, the ship will never return. You're never going to see land nor strand anymore.

Well, it was the topic of discussion on the whole trip, and soon I heard one man say, "By God, this one won't come back no more." Then another said, "Boy, don't worry about it. There's only another couple days left." I tried to convince the worried man that it was all nonsense, but he was one chap who could never be convinced. If he could have swam ashore himself, he would have followed the rat. One of the fellows joked to this man who was so concerned. He knew we were going home that same day, after we got our gear in, and, just before they left the vessel to take in the gear, he hollered out, "Mike, b'y, nothing doing with old bugger rat this time. We'll be home 'fore daylight."

Strange, I was never superstitious when I was fishing on the banks, but I've heard talk of fellows from Grand Bank, when they'd shape up their course for home, they'd swing right around, come right around with the sun. If you go against the sun, they'd say, it's unlucky, but it never made any difference to me.

The Dutchmen, the Lunenburg skippers, were the superstitious ones. If we were on the banks and happened to come across a Lunenburg schooner, we'd just heave up, go aboard and pay a visit. If one of our men went back to talk to the captain, a couple of the men who carried him over in the dory would go to look down in the hold. Most wouldn't mind that. But if the skipper saw you do that, he'd almost heave up and go around. They called it unlucky.

Our dories had about all of their gear in, and I went along and spoke to the man with the outside dory, about one or one-and-a-half miles from the other side dory. It was Uncle Bill, the same man who was washed overboard in the August hurricane on the previous trip down on the Grand Banks. I called out to him and asked, "Uncle Bill, how many lines do you have left?" He said, "Two lines, sir. That's all," which meant all the other dories had approximately the same left. Well, that was fine. Then I took the wheel and heaved down to get the schooner headway enough on the other tack. With sails, no engine, just give her enough headway to go a berth from that dory (about sixty to eighty meters). And go around, and by that time, he would have his gear in, and the next dory too. And I'd pick up all the dories as I went.

I took the wheel and hove it around, and when she came into the wind and sea, she went up and down. Right head to the wind, the sails were flattened because she was so fair. She gave a plunge, and I heard a crack forward, and realized the jib stay on the bowsprit had broken off from the masthead. Then she gave another plunge, and the jumbo stay broke. Well, I knew what was going to happen then.

The kedgie, the young fellow who was deckhand with us, was standing alongside me at the wheel. I grabbed him by the sleeve and we jumped for the cabin. By the time we got on the cabin floor, I heard the two eighty-foot spars crash right fair back over the cabin gangway, and back over the stern.

The mast was so fair over the gangway that there was just room for the two of us to get up through. Before we came up through, with every pitch she made, I heard the water gushing in through the stern where the masts broke the stern of the vessel off completely. And it was coming in through the cabin. What a mess we were in. The sails, the mains'l and fores'l, were out over the side of her. I looked forward. By this time the cook was up forward hollering out, "Was anyone hurt?" But we weren't hurt. If it had happened two nights before, when we had to take in sail, coming in from the Grand Banks, there would have been another big disaster.

There were our ten dories, all of the men were out. The last dory was almost three miles from us. It was our habit to look at the mothership now and again, and, one after the other, they glanced around. By that time some of the dories were coming aboard. They all came aboard within an hour or so, and we saved our clothes and got off again before she might sink.

God, skipper, there's something down there like a slab on the water, but there's no masts

There was another schooner fishing about three or four miles from us all day. We could see her. It was a fine day. Captain Orlando Lace had his gear in before we did.[2] He was fresh fishing from Lunenburg, but he was going to North Sydney. And after he had his gear back and dories and men aboard, he went in the focs'l and had his supper. He was going to land his trip. When he was finished and walking back from the focs'l, back to the cabin, he said he was sucking his teeth and glancing around, and he looked back and asked, "Where the hell is that vessel that was fishing down there all day? We're not out of sight of that vessel yet." He knew we couldn't sail out of sight as quick as that. Then he got curious. One of his men went up the rigging to the masthead, looking, and he said, "God, skipper, there's something down there like a slab on the water, but there's no masts. I don't know whether to say it's a schooner or not, but there's something down there where that vessel was fishing all day." So he grabbed the wheel from one of the men to swing around on, to come back. He had an engine in her, so he could steam about ten miles an hour.

But his men said, "My God, Skipper, you're not going down there?" The war was on and they said, "That's a submarine. Submarine got him sunk. That's what that is." "Well," he said, "it makes no difference. We'll go back and see the poor bugger and see what's become of him, or what's going to become of him." Some of the men didn't think it was right to go back if the submarine was there. The vessel's name was the *Mahaska*, and Captain Lace was a native from Rencontre East, Fortune Bay, but now lived at Lunenburg. He was a First World War veteran. The captain took the wheel himself and changed course towards us without losing any time. Anyway, by the time he got down to us, most of our men had their belongings in the dories, and she was sinking. He got us all aboard, took our dories in and put them on his deck. He had ten dories and we had ten, so there were twenty dories on board.

After getting all of the men on board, I took the cask of gasoline on deck that we had for the deck engine, and the mate and I spread it down below, through the vessel, and set fire to it. She sank within two or three hours. There was nothing else we could do. That's the last thing you do. When you see it would be a menace to navigation, you have to set fire to your ship. 'Twas some hard to look at the *Florence* go down, one of the hardest things anyone could ever look at after sailing in her. But she went down within a short time. She went down with her head up high, stern first. She was a nice vessel. A flash went through my mind, and I said to myself, "How lucky we were that she brought us through that critical time in August month."

Now we would have been home about two o'clock that morning. But it took another eight days. That night a gale of wind came up when we were going in for Sydney in this schooner. We would have had it bad in the dories. Some would have capsized because it was blowing hard.

We got in Sydney late the next evening, and spent four days in a hotel. Then we were transferred to Halifax, where a boat carried us down to Grand Bank. Most of the men didn't come to Grand Bank. Only about half a dozen lived there. Hardly any were what you'd call "real" Grand Bankers, because our men used to come from Fortune Bay.

That was the last deep sea fishing **vessel** in Newfoundland. As I said earlier, the owners were going to put an engine in her when we got home that fall. All the other schooners already had auxiliary engines, and it hasn't been easy by any means during the last couple of years to try to keep up with them. But we were fortunate and I was well pleased with the amount of fish we landed. We had landed a pretty good voyage that year. We had about 1500 quintals in her then. That was the last.

We were about 120 miles from Grand Bank, and sometime that winter the main mast of the *Florence* drove right in down below Grand Bank. So that was the last of her.

When we went in Sydney, I noted a protest[3], wired a telegram to my owners, and one to my wife. And the strange thing about it, my wife got that telegram five or six hours before the owners got theirs. Naturally, I had the same on her telegram as I had on the owners': something like "Sorry we lost our vessel." And, of course, some of the children were in the house when their mother got the telegram, and they went right out and said, "Dad lost his vessel."

They never gave me a very good reception

I don't think the owners were worried about the fees involved, but they were upset to think that I should wire my wife first, and not wire them. I tried to convince them, but they wouldn't believe me. But I really did wire them, on purpose, at least four or five hours before I wired my wife. And I figured that, in five or six hours, they'd have their telegram. But they didn't. That was all. The company didn't have another vessel for me and didn't say they were going to try to get me one or anything like that. The way the times were getting — they weren't doing well on fish prices — maybe the owners were glad the vessel was gone. But some members of the firm, and one in particular, got kind of nasty because I did that. They never gave me a very good reception. But I didn't mind it, I laughed at them, because I was still young (forty-one). But I was without a schooner then.

We were insured against that fish. When you go out on a trip in one of the fishing vessels, your 'prospective catch' would be insured. Even if you didn't have any fish aboard, you'd get paid for about $5,000 worth. It would depend on the fish price. We had about 500 quintals, so we had that much towards the insurance.

The owners got about $5,000, prospective catch, on that particular trip. It was divided up just the same as if we'd landed a trip of fish worth $5,000. The only exception was, the men normally paid for half of the fish making. You'd wash out your fish and put it on the beach. Then the women would make the fish; spread it, take it up and everything like that. They'd get so much a quintal for it. The rule was, the owners paid for half of that fish making and the crew the other half. But none had to come out of that $5,000 like that because the insurance took care of it. So, we had our voyage landed and the men never lost anything they might have made on the voyage. The owners lost the vessel, of course, but she was insured. And now I had to look around for a job. So I went home and got settled away.

Chapter Nine

In Another Dummy Amongst the Big Power Vessels

In late 1940 Arch is once again a captain without a vessel. In short order, however, he was offered command of the Lunenburg-owned *Pan American*, another dummy schooner. If he reflected on his disappointment thirteen years before when his first command was unexpectedly sold away by its Lunenburg owners, he was quick to accept. Even if it was another dummy, he felt he was "moving up" in a better regional industry, one already involved in the lucrative fresh fish industry. He and his crew would gain by a vessel well founded and supplied by its owners, superior work conditions and income prospects, more accessible and frequent accounting, and the cooperative relationships these differences fostered.

The move wasn't all rosy. The *Pan American*'s new captain faced great competition for labour from military service and expanded wartime wage labour construction opportunities ashore, building and servicing military bases in Newfoundland and Labrador. And the war increased the hazardous conditions banks fishers faced every day. To get fishing crews, whether for dummy or powered vessels, the average crew share was enhanced with wage guarantees. But even then it was difficult.

Engines had appeared in New England and Nova Scotian schooners before the 1920s. The *Pan American* received an engine in January 1942. Skipper Arch recalls that Lunenburg auxiliary power vessels had bigger engines than the "little sewing machines" in Newfoundland schooners. Banks fishers and vessel owners knew that auxiliary power enhanced operational freedom, catches, earnings, and profits. If Newfoundland vessel operators were slower to adopt power, it was not for romantic attachment to grit, muscle and sailpower. Marine engines were scarce in wartime.

Newfoundland fishing interests began expansion to offshore banks fishing in the early 1880s. It was in part a response to the competitive example of American and other foreign banks fishing. It promised larger, more reliable landings and profits. The resource was open to all takers, and perhaps there was fear that foreign competition would eventually affect nearer stocks. This expansion involved acquisition of ever larger, more mobile, and eventually mechanized vessels. These changes reflect an accelerating intensification of fishing. As we will see later, it soon led to revolutionary changes in Newfoundland's fishing industry and communities: full-blown industrial fishing, with groundfish (otter) trawling, fresh fish processing plants, and year-round fishing and processing employment in a cash economy. The process culminates in annihilation of Northwest Atlantic fish stocks with vast new implications for Newfoundland's industry and people.

1940-1942

The news soon spread around about the loss of our vessel, and I was home only a short time when a telegram arrived from a party in St. John's who asked me to come down and look over a large, three-mast, French war prize vessel. It had been towed into St. John's after the Vichy Government took over in France. Wherever there were French ships, then, Britain captured them. This party had the vessel and wanted me to examine her and give my opinion about converting her into a deep sea fishing vessel, to fish from Greenland. I went down to St. John's and looked it over, and I didn't like it. It would cost too much money to make her suitable, and I advised them of all that had to be done. A meeting was held and the expected expenses tallied up, and they agreed it was too costly. The company was nice, and paid me well for the trip, which took about ten days.

When I arrived home from St. John's there was a letter from Lunenburg, Nova Scotia, waiting for me. A company there, owned by a Senator William Duff,[1] a man from Carbonear, Newfoundland, heard I lost my vessel, and they asked me to take one called the *Pan American*. She was the last dummy deep sea fishing vessel operating from Lunenburg. I knew the vessel, of course, and had seen her many times. Although it was another sailing vessel, I knew quite well it wouldn't be easy to get a vessel with power, because all of them already had their captains. It was just as well to keep on struggling, so I accepted the offer.

It was nearing Christmas by then, and I had a lot of work ahead of me for the next couple of months, to select twenty-five men — an eleven dory crew, cook and deckhand — and go to Lunenburg, take on supplies, and bring her to Newfoundland to take on my fishing crew. It was the same routine as when I first took charge of the *Vera P. Thornhill*, my very first command. But even if it's another dummy, I felt more secure and had more confidence in myself. I knew I would have to struggle hard, especially going to Nova Scotia, where the power vessels were really powerful. By comparison, the little engines in our Newfoundland fishing vessels were like little sewing machines.

When I accepted, to go master of her, I started shipping my men, and by Christmas I really surprised myself by having a full crew, and all good men, including one of the best cooks in Fortune Bay. I managed to have January month for myself with my family and not much to do, except for feeling around among the different companies to find out which would be the best one to be my agents for the coming season. It meant a lot of business and I could pick my choice. All of our men had to get their supplies for themselves and their families for our first spring trip, and the vessel had to have a lot of supplies and gear during the year.

I really looked forward to the coming season, and I had been promoted by going to Lunenburg — even in a dummy amongst the big power vessels. I knew quite a few people there and in La Have. I had the feeling before I started fishing this year, no more dummies. My record for the past four years was good, and a big help to me. I didn't expect wonders, but the business people up there are reasonable people, and they knew quite well one cannot land as much fish without power as one can with power.

If only we could afford to get a piano

By now our two little girls had been going to school for a few years, our oldest would be eleven on her birthday and the youngest eight. They are both very musical, and a number of times my wife and I said to each other, "If only we could afford to get a piano." But the times had been so bad since we married, it was impossible to get one. If we had the interest that we paid on our $300 loan for many years, we could have afforded one, but that was out of the question.

There's no mistake about it, during the first three or four years of our marriage, it was really bad for everyone. I remember when our first child was about to celebrate her second Christmas. We went to a store on Christmas eve and bought a small rocking chair for her. We had one $10 bill to our name, and the chair was five dollars. We bought that chair and both of the girls got a lot of pleasure out of it. Many people then were worse off than we were.

My wife had a foot sewing machine when she got married, and, as our two girls got old enough, every now and again they would go to it and pretend they were playing the piano. About a week before I left to go to Lunenburg I missed them one evening and went to the room where the foot machine was, and both of them were taking turns pretending they were playing a piano. I said to my wife, "If only I can get the amount of fish I'd like to get, about 1,700 quintals, on our spring trip, and there's a secondhand piano at Lunenburg, I'll have it when I come home this spring."

The first week in February I took on five men, a skeleton crew, and we joined the S.S. *Baccalieu* at Grand Bank and went directly to Halifax, and from there to Lunenburg, where we joined our vessel, *Pan American*, and immediately began making preparations to fit her up and get her ready for fishing. After being at Lunenburg for a few days, I put an ad in the paper for a piano, and within a very short time I got a call and went to the person's house. No mistake about it. It was the right piano, but the person was asking $30 dollars more than I was planning to pay. And, besides, I still had to fish my spring trip, so I told the man I thought it was a little too much and kept my fingers crossed and left.

It took a week at Lunenburg to get everything ready. And there is no mistake about it, this company really wanted everything to be ready. We got the vessel's quarters painted and varnished and all our sails, etc., put in first-class condition, took on supplies and salt, and food like I never saw put on board a vessel since I left the *Vera P. Thornhill*!

Finally, the company put the very best linoleum on the cabin floor that could be put on any new vessel. They also put a new mattress and rugs and pillows, etc., in my berth. I can assure you, when we arrived at Grand Bank, I wasn't ashamed for anyone to come on board. Although a dummy, the vessel meant a lot to me. What a contrast to what I had seen in all my years fishing, with the exception of 1927 and 1928!

We were back in Newfoundland on time to start fishing with the rest of the vessels. The first week in March I had all my crew on board and sailed for the banks. I decided to go on the Western Banks, on Sable Island Banks. If we had good luck and got a good trip, we would go to our home port, Lunenburg, and land the trip. But it didn't work as I had planned.

God, skipper, there's a submarine comin'!

We went up on the Western Banks, on Sable Island Banks, and a strange thing happened there one day. We were just anchored, but there was too much wind to fish, and all the men were around. By and by someone looked down and said, "God, skipper, there's a submarine comin'." I said, "A submarine?" "Yes," he said, "a submarine." I could see her myself then. It steamed right up, first with just her periscope out of the water, and then came right up. She was washing pretty good, because there was a big breeze of wind. It came right up alongside. When I saw her coming, I put out our flag to the top of the mast.

It came right up alongside and he called out and he asked me a few questions: Where I was from? I told him I was from Lunenburg, Nova Scotia. How long had I been out? What time did I think about going in again? And then he wanted to know how I was set for supplies. When he came to that, of course, then he began to make everybody feel kind of happy. We didn't know what he was, don't know yet. "Oh," I said, "we got plenty, thanks. We got plenty of supplies. We haven't been out very long."

"Oh," he said, "put out one of your boats and come aboard and we'll give you some biscuits, cocoa and soup, and some different things."

We couldn't very well turn it down, so we put a dory over the side and two men and I went alongside. They passed five or six cases of stuff from the deck of the submarine. They talked like Englishmen, and were very good to us. When we went to leave, we wished them good-bye, and I said, "We thought you were Germans. We were a bit scared, you know." And they laughed. But I don't know yet what she was: she could have been an enemy submarine, because they could have had someone who spoke English. And there were no identification marks on her. Of course, that didn't matter much then, anyway. They didn't do us any harm.[2]

We went on fishing, kept on going. But we had a poor trip that time, and only got about 250 quintals. Still another hard struggle. The fish were scarce and the weather stormy, so we didn't have very good luck on our first baiting. We returned to Grand Bank and took on supplies again, and went in Fortune Bay for another baiting. Most of our vessels were in and gone again on their second baiting, and most of them were well fished. I learned they were all headed for the Western Banks, in the Gulf, and to Rose Blanche Bank.

Tis hit or miss. We're goin'

'Twasn't very late in March and, as a rule, our vessels very seldom go to the Grand Banks in March. But the *Pan American* was going there this time. We took the chance because of the bad baiting on the Western Banks. "'Tis hit or miss," I said. "We're going." After getting our bait on board, I touched in at Grand Bank on my way along for a supply of food, and advised my agents to wire our owners at Lunenburg and tell them we only had 250 quintals on our frozen baiting and had left again on another baiting. I didn't tell them where I would be going.

When we got out to the entrance of Fortune Bay we shaped our course for the Grand Banks, with a strong northwest wind in our favour. All that night and the next day we made good time and the following morning, after coming over thirty hours from our departure, about 250 miles, we were to our destination. It looked very lonely down there this time of year. Not a ship in sight.

We began fishing immediately after anchoring, and there were a lot of fish there. In no time at all after the men went to haul their gear, I saw the dories coming, loaded. Eight days from the time we anchored, we were on our way back again with 750 quintals on this baiting. And now we have a total of 1,000 quintals on board, which is marvelous for a poor dummy.

Hands dressing fish aboard the *Pan American*, c.1940.

One night while we were anchored there, there was another scare. We saw something coming towards us. Just one light, that's all. And you could see it a long way off. By and by, it seemed to be coming near, but it was so long before it got to us, it must have been two or three hours. As it got nearer it seemed more powerful. She came right up alongside and we couldn't tell if it was another submarine. That's what we were scared of. But it wasn't. It was a Lunenburg schooner, one of the large power vessels down from Lunenburg. He happened to come right down where we were. He came along and they hollered out and asked, "How was the fish?" And I said, "There's a fine lot of fish here. We've been here four days now and have 400 quintals."

So he went just a berth from us, about four miles off, just so the two ends of his gear and our gear wouldn't meet together. That's what you call a good berth. They also got themselves a large baiting. We fished there for three more days, but there were two or three when we couldn't fish because of the weather. But we got about 1,000 quintals aboard now. Up stakes, again, then, and now we're feeling a little better.

We quickly got another baiting in Fortune Bay, and I wired my owners and told them the news. We went off on the Grand Banks again, and on the twentieth of May we arrived at Lunenburg with 2,000 quintals for our trip. I wasn't too excited over it, but I was pleased and so were my owners.

I didn't realize what a trip we had until one day while we were discharging. The manager sent for me, to come to his office. He passed me a newspaper from Gloucester, U.S.A., and it had a big writeup about how the last and only dummy schooner sailing from Lunenburg, Nova Scotia, commanded by a Newfoundland captain and crew, arrived from the Grand Banks off her spring trip with 2,000 quintals of fish on board. Our manager, Senator Duff, had been living in Nova Scotia for years. He was more excited than we were. We were well up amongst the highliners with the power vessels there and in Newfoundland. So we had a pretty good summer.

We all knew by now that we were going to make a good spring trip, and receive a much better price than we could get in Newfoundland, and a statement for everything, from a needle to an anchor. I ordered the piano that I had seen earlier and the company crated it up for me.

Within five or six days, we were on our way to Newfoundland again with full supplies. And our crew all had supplies to take to their homes. After arriving at Grand Bank, I gave the crew a few days home with their families before we took on board our caplin baiting. As quick as we arrived in Grand Bank, which was early in the morning, before the children went to school, fourteen or fifteen men got around the piano, hoisted it out of the hold of the vessel, and carried it to our home. Our family didn't know we were near when one of the little girls looked through the window and said, "Oh, mom, look! The men are coming with a piano!" Our youngest daughter was only eight years of age, she took off somewhere after she came out of school in the afternoon.

When she came home, she said she had a music teacher. The two of them went on from that.

After a few days at home, we got another caplin baiting almost in Grand Bank harbour and left immediately for the banks. I was very pleased with the success we had on our caplin baitings. We fished two squid baitings after that and went to our home port, Lunenburg, around the 20th of September. We had had another big trip, and our owners were very pleased, as we were, and they informed me that they were putting a secondhand engine in our vessel. It was a heavy duty engine, seventy-five horsepower, that had been taken out of another vessel. Needless to say, I was happy and glad to have it, even if it was secondhand. I knew our company was aware of what they were getting.

When our vessel was cleaned up and all of the fishing gear and dories, etc., put ashore, the men settled up. It was brought on board the vessel and every man could see for himself every cent of expenses that had to come out of his share for the trip. This was far different from the settlements we were used to getting in Newfoundland. This settlement was for our spring trip, and the settlement of the trip just landed would be sent to every man before Christmas. The crew got all of their money from the spring trip, and the company advanced each man approximately half of the amount made on the last.

The men, my crew, came down to Halifax by train, and came home on the steamer, while I stayed behind. There were quite a few things I had to see to regarding the engine and its installation before I left Lunenburg, but there was a delay, and the owners gave me permission to come home. They would inform me when things were finished. We decided before I left for home that I would get an early start in the New Year, and go winter fishing in the Rose Blanche area.

It was getting very dangerous on the fishing banks . . .

There was some unrest amongst both the Newfoundland and Nova Scotia deep sea fishermen during the past season. The

German submarines sank quite a few merchant ships right on the grounds where the vessels were fishing, and some of the fishing vessels had been sunk by convoys in dense, foggy weather, especially at night. It was getting very dangerous on the fishing banks and I doubted very much if all of the fishing schooners would get sufficient crews for the coming season. Some of our Newfoundland vessels have already had to abandon their voyages and tie up because of men leaving.

And there were other things that kept men from going deep sea fishing. Many of our Newfoundland fishermen were working at Argentia, on the American base, and on other bases. Some Newfoundland vessels did not get crews to engage in the fishery at all this year. I had been one of the fortunate ones. I never heard one of my men murmur for the year, even on the old dummy again. Even though I had had to struggle a lot during the past three years, I still had a lot of pleasant memories of the old sailing ships. But I hoped I was through with the dummies.

I enjoyed my stay at home with my family since the last week in November. It was quite a change for me, as I had been on the go nearly every fall and winter since we were married. By Christmas I had another full crew for the coming season.

I was with the big shots

Around the middle of January, 1942, I took a skeleton crew of five men to Lunenburg, took on board all our fishing gear and supplies, food and salt, etc., and arrived back at Grand Bank during the first week in February. I was very pleased with the secondhand engine, and everything worked well. The vessel could steam about six or seven miles per hour, which was the speed all of our Newfoundland vessels made with the light duty engines. 'Twas a lot for me. I was well away then. I was with the big shots.

By the middle of February I had all my crew, eleven dories, and bait on board, and we were on our way. You almost always have some changes each year, but I had over

A view of the foredeck as the *Pan American* creeps through dense fog, c.1940.

half of the men I had the year before. Quite a few of our Newfoundland vessels couldn't get any men to go fishing this year, and most that did had to offer wage guarantees before they sailed at all.

Rose Blanche proper was our official harbour for the next month or so. I said a few things about winter fishing earlier, but there were no engines then. The Western Shore is a hard, rugged coast in the winter, and you expected to be caught out in some storms, as we had to go and come twenty miles from Rose Blanche harbour. There was another vessel fishing there from our firm at Lunenburg; her name, *Eileen C. MacDonald*, Captain Edward Cleveland. It was practically new and had large power. Her normal speed, ten miles per hour, wasn't so much. But it was a big help.

Winter fishing on the Western Shore that year was so hard on our engine, she only made about three miles an hour after that. And two skippers, Ab' Myles and Sidney Harris, used to kid me about how fast we were steaming. They could steam twice as fast. We were great friends. One day we were standing together on the pier in Grand Bank, and Sid' was telling another fellow, "Boy, you should have been in Lunenburg the other day. Arch was steaming down to the dock, she was bubbling along and the foam coming from the back!" And I said to both of them, "You can kid all you like, but I'll have something, someday, with so much power, I'll tow you fellows faster broadside than you ever steamed!"

Years later, when I was fishing off St. Pierre in the *Blue Foam*, one day I saw a schooner about two or three miles away with her dories out and I heard her skipper talking on the radio set. It was Ab' Myles with the *Marshall Frank*. I heard him say, "I see our friend, Skipper Arch, he don't have to bother no dory fishing. He's there alongside us towing his net, you know."

I never said a word over the set. We took our gear in with a net full of fish, got it all iced in on deck, and I called out to the men, "Now hoist in your doors." They looked up and said, "What in the name of God? Has the old man gone crazy? What's he going to do, hoist in the doors right in a smack calm?"

I gave the engineer two rings: Emergency! Give us all the power you have down there! She could clip off about twelve or thirteen miles an hour. I steamed up and went right 'round him. Coming back for the second time, I turned the set on when we were abreast and he said, "You S.O.B. You never forgot it, did ya!?" Well, how much we laughed over that.

After arriving at Rose Blanche, we met stormy weather for a few days, but there was no talk of ice conditions, which we are all concerned about in this area at wintertime. In our first day on the fishing ground, we secured 100 quintals. What a blessing to have power!

While we were on our winter trip, we got thirteen fishing days and secured 1,250 quintals, which was excellent fishing, especially at wintertime. It wasn't thirteen days in succession. In wintertime, you had some bad days. The weather had to be watched very carefully and we were careful not to miss the forecasts in each radio broadcast.

One day we got caught and couldn't make it into port. We left Rose Blanche at around two o'clock in the morning, and were on the fishing grounds waiting for daylight to come. It wasn't a good looking morning by any means, but I decided we would set half of our gear. We didn't lose any time getting it back, and there were quite a few fish on the gear. We left for Rose Blanche around midday. When I was taking my last dory on board, snow began to come on us. By three o'clock in the afternoon it was almost gale force. We kept going, hoping we would make it, and, finally, we had to heave to and dodge out to sea again.

It turned out to be one of the worst storms that winter. The Gulf of St. Lawrence is a hard place to ride out a storm such as this one, in the mid-winter months. At first, the wind was from the southeast, when we were trying to get in, and it stayed that way all night. The next morning our barometer had fallen almost as low as it could, and towards daylight it came from the northwest, and the following night it was below zero. Our men had to be on deck most of the night to keep the ice off the vessel as much as possible, and some got their faces and hands frostbitten. The storm lasted quite a while, and we never made port for almost three days afterwards. We were loaded with ice from the mastheads to the bottom of the vessel.

Since power came in, all the bank fishermen had pilot houses (also 'wheelhouse') on their vessels. In the late years, when a new vessel was built, they added on the pilot house. But on vessels that were already sailing and converted to power, the pilot house was put on later. It was a tremendous improvement. Before the pilot house came along, the man steering the ship was exposed to all kinds of weather, summer and winter, so it's not hard to understand how much cold and frost men have suffered with winter fishing, especially in the Rose Blanche area. And many times men had to be lashed to the wheel (with the 'wheel rope') in order to keep from being washed overboard when a sea struck. The pilot house was built over the wheel of the vessel and over the gangway going in the cabin, and probably seven or eight feet high. The invention was surely appreciated by all fishermen.

A pilot house could not be put on the *Pan American*, however, because it lacked sufficient power. We still had to carry our mainsail. The mainsail wasn't used when fishing, but it was used on long passages. As mentioned earlier, another vessel from our company was fishing here with us, and, being practically new, it had a nice, comfortable pilot house.

I decided to try a new stunt on the first stormy day....

There are always spare sails carried on a vessel, and we had a spare old mains'l. Our intention was to take it on the trip with us to cut it up in pieces for men to kneel on when salting fish in the hold. The first two or three days we were fishing it was very cold and frosty, so my crew and I envied our friends and the captain on our sister ship with the comfort they enjoyed with their pilot house. I began thinking, but not saying a word. I decided to try a new stunt on the first stormy day that came along, and you didn't have to wait long.

The day came and we took our old mainsail out of the sail loft, got it on deck, and I sent two men ashore to buy some wood framing. We had all sorts of carpenters on board. Every man on board was very interested and wondered what invention the old man would think up next. When the framing was up, the old mainsail, a very heavy canvas, was cut up and a real pilot house made, with all the necessary windows. I think everyone else around thought we were crazy when we were building it, but it suited the purpose fine. It was much warmer, even in the cabin. It was frosty that night and all the next day, and no one went out to the fishing grounds.

The next morning it was suitable for fishing again. It was cold and frosty, but I can't express how comfortable it was going out on the fishing ground. The heat flowed up from the cabin and men were in their shirt sleeves, steering the vessel along. Everybody really enjoyed the pilot house during the first two nights, but on the third morning we shipped a sea that took the whole works away. That night we got back to Rose Blanche at about ten o'clock, and Captain Cleveland was on his deck while we were passing by, and he called out with his Lunenburg accent, "Where is your pilot house, Awch?"

In early April, we proceeded to Lunenburg with a good trip, and weighed out 1,250 quintals for the thirteen days of fishing. We averaged almost 100 quintals a day. Pretty good. We were all very pleased, and, to make things better, the price of fish was much higher that year.

We all worked hard that winter and it was very cold. The old engine worked hard too, and had to be repaired while we were discharging our fish. Then we came down in Fortune Bay to give the men a few days at home. Within a week we were on our way for another trip around the last week in April. We were after big "trout" that spring, and fished two baitings on the Grand Banks and secured another 1100 quintals, so now we had 2350 quintals since we started the winter. There was every possibility of being one of the highliners again that spring.

Skipper, our rudder is drivin' astern!

We had time for another baiting, and, after we secured it, we arrived at the same position on the Grand Banks where we fished our two previous baitings. We knew there was a fine lot of fish there, and there were other vessels around. 'Twas a nice morning. All the men were out setting their gear, and I was in the focs'l having breakfast, when my brother-in-law, the engineer, looked down through and said, "Skipper, our rudder is drivin' astern!" The rudder came off, and here we were out on a fresh baiting with 1000 quintals aboard!

We had a spare dory on board, so the engineer and I put it out and towed the rudder alongside and tied it on until the men came. Then it was hoisted in on deck. Then I sent all the men to take in their gear, because there was nothing else to do. Now we had to try to rig up a jury rudder and get the vessel home, because you weren't allowed to fish away like that. If I had continued to fish when I knew the rudder was gone, even when I knew the vessel would handle, and anything happened before we got to Lunenburg, the insurance company would have said, "Well, you lost your rudder a week before that." It would be bad. It was pretty sad to be out on a fresh baiting, knowing the fishing was good, and have to abandon our trip. Of course, we still had a good trip, but we would have had much more but for this incident.

Another vessel from Grand Bank was just a berth from us, and I went aboard for about fifteen or twenty minutes, and told him our troubles. And I said, "we have to leave and go all the way home. There's nothing else to it." When I got aboard, we put the mains'l on her, because the engine was small. "We won't rig up a jury rudder yet." We just hove up our anchor, put the riding sail, fores'l, and jumbo on her, and started the engine. I was anxious to see how she was going to handle herself. It's almost too much to believe, but when I hove down on the wheel, she worked pretty good. If that rudder would have come off, gone astern before daylight, we wouldn't have known about it. We sailed from there, about 250 or 300 miles down on the Grand Banks, right up to Lunenburg with just a rudder post!

We were doing fine all the way along until we hit a dense fog and ran up on a reef about forty miles from Lunenburg.

We just about lost our vessel, with a little over a thousand quintals on board. But I rang down for the engineer to give us the bit of speed we had: "Full speed astern!", which wasn't very much after using that little engine hard that winter. But our little power backed the vessel off the rocks, and she floated, but leaked badly. We got to the pumps and managed to get home safe, discharged our fish, then docked the vessel and put the same rudder on, and left again within eight days for our caplin trip. We arrived home the twentieth of September with another good trip. After it was discharged and all the necessary things at the end of a voyage were done, our men all returned home with a very good paying voyage.

Many Lunenburg fishing vessels went to the West Indies with a load of drum fish, and our owners decided to send the *Pan American*, and they asked me to take her there. I decided I would love to make another trip to the West Indies. I hadn't been there since the trip to Barbados in the schooner *R.L. Borden*. The war was raging, but, heck, there was no more danger going there than fishing on the Grand Banks those days. Some of the Lunenburg vessels had to abandon their voyage after their spring trips because men would not sail, as there were so many ships getting sunk on the fishing grounds.

The company hired on a few local men to do the cleaning and painting necessary after a fishing voyage. I stayed at Lunenburg to supervise the work to be done for the West Indies trip. It took some time before the vessel was all put in order, and the necessary sails sent to the sail-loft for repairs, etc. And then the company found it would take longer than expected before the fish cargo would be ready. This meant the vessel would be later getting home again. We had already made plans to go winter fishing again at the same time as last winter, so the owners suggested I go and have a few months at home, and select a crew for fishing, in order to be ready by the time the vessel returned from her trip.

I arrived home during the last week in October, fully intending to be with my family for a couple of months, and I told them, "Now, then, I'm going to be home until January, anyway, because we're going winter fishing again. I'm home for the winter." "Good!" They were glad to hear that. I had been out to sea since last January.

One night, about a week or two later, there was a knock on the door. It was Charlie Tibbo, a businessman in the Forward and Tibbo store in Grand Bank. He told me he had a call from Steers Brothers, down in St. John's. They heard I was home for the winter months, and they urgently needed a mate on the *Henry Stone*, one of their freighters. The skipper, Captain Ben Snook, belonged to Grand Bank. Well, they offered me $100 a month and I said, "No. I'm not going. I'm home for the winter and I'm going to stay."

I don't think the family cared too much about the idea — the children were small. A skipper got $120 a month then. But I went on and said, "You tell Steers, if they'll give me $150 a month and pay my way to and from St. John's, I'll think it over." I didn't think they'd accept that, but the next morning, almost before I was out of bed, Charlie was up again. The same firm had called, explaining that they understood my situation regarding being master of another vessel and that I would be going winter fishing in January. But they wanted me to accept, and they would transport me home when need be, no matter where the ship was. I came down and joined the vessel then, and ended up with about $30 a month more than the captain. So I was mate for Ben after skippering for a good many years.

We went up to Sheet Harbour, Nova Scotia, loaded shingles, then discharged it up the Quebec coast. Then to Montreal for a load of cargo, and brought it to St. John's. I was on the *Henry Stone* for two months, and arrived home two days before Christmas. A few days later a letter arrived from my company at Lunenburg, explaining that the *Pan American* was sold. They found an excellent market for their vessel and decided to sell her down in the West Indies.

Well, then, I was out another schooner. They say every cloud has a silver lining. As I found out later, the silver was just beginning to break through for me. This was the best thing that ever happened in my career.

Chapter Ten

A Wonder for Her Time

Captain Thornhill's story frequently portrays Newfoundland's banks fishing industry as less progressive than Nova Scotia's. Thus south coast schooner firms were slower to install engines in their schooners, and when they did, they tended to be smaller, less powerful. Moreover, Newfoundland banks fishing firms kept more exclusive ownership control over vessels, sustained outmoded dominance over fishing labour, supplied their schooners meagerly, and when it came to settling accounts with fishermen how they kept "the books" probably reinforced long-standing fisher suspicion of merchant dishonesty and exploitation. In short, elements of a system of strong merchant — and credit — dominance of fishers endured into the second World War years. But the industry was changing in 1943, when Arch joined the Harbour Buffett firm of W.W. Wareham's, which evidently chose to follow practices of their Lunenburg counterparts. Did maritime labour scarcity influence them?

The full story of Newfoundland's banks salt fish industry reflects important developments that intensified fishing. This occurred within the framework of a credit driven economy dominated by fish merchant-traders. Its early success hinged greatly upon cheap, able and obedient labour. That resource became less available during the 1930s and even more so in World War II. However, the saltfish industry's decline is complex and its full story is best told elsewhere.

As seen here, wartime brought opportunities for some of Newfoundland's fish export industry to supply lucrative fresh fish markets. These markets for a time enabled cash settlements with banks dory fishing crews, trip after trip. Aboard the *Ronald George*, Captain Thornhill and his crew of banks dory fishermen found themselves switching from salt fishing, to fresh fishing, and then back to salt fishing at war's end. The firms appear ineluctably drawn by profits to be made supplying growing and more accessible post war fresh fish markets by adoption of the "new" groundfish trawling technology.

During the war and through 1947 Arch and the *Ronald George* continue the old fishing voyage and winter freighting annual cycle. But they now have engine power that, together with landing fresh fish in return for cash settlements, trip after trip, gave new meaning to the idea of "making hay" while opportunity permits. Banks fishing took on a new, frenzied intensity. It is a harbinger of what would follow in 1948.

Equipping the *Ronald George* with the first two-way radio communication on Newfoundland schooners in 1945 reflects management's desire to strengthen the link between catching vessels and shore processing operations. Perhaps the Warehams even thought radio communication would improve their schooner fleet's fish production. These radios no doubt enriched communication between skippers aboard their vessels at sea. Yet skipper Arch disavows their value as far as information about where the fish were because skippers lied about their catch. He links this practice with recollections of inter-skipper information management 'stunts' among schooner fishing skippers even before two-way radio transmission. We may wonder, of course, if this is the full story. Messages that reveal only a position among limited possibilities may have value.

In this chapter we also see how the countless events and bits of information gathered over a lifetime are sometimes truncated and overlapped. Hence Captain Thornhill refers to the Newfoundland Dockyard as a C.N. and C.N.R. facility, designations that became true upon Confederation with Canada in 1949. Arch penciled most of this chapter's information in a scribbler. Its order and details suggest he based it upon a record of some kind, perhaps a private file or collection of telegrams I never saw.

1943-47

Shortly after the New Year 1943 I got a call from another company in St. John's asking me to consider relieving one of their captains for a month, to go freighting between Newfoundland and Halifax. I accepted the offer, and the boat, the *Wymota*, called for me at Grand Bank. My first trip was to go to Isle aux Morts on the southwest coast to take a load of fish for Halifax. It turned out to be one of the most blustery and frostiest winters in the history of Newfoundland. From Halifax we went back on the southwest coast, and the company asked me to carry on for two months. On my second trip to Halifax we encountered a lot of heavy Gulf ice and had to come down from Rose Blanche to St. Pierre and Miquelon to get around. And on my birthday, February 27, we encountered a severe storm and ran into heavy ice and had to run to the southeast to get to the south of it. In Halifax we discharged our fish, filled the ship's hold with cargo and went to the south side for a deck load of barrel (fuel) oil. Our vessel was pretty deep in the water.

On our way to St. John's we ran into another storm on Banquero, our deck load got loose, we lost a few barrels, and the men and I were on deck most of the day up to our waists in water, trying to secure the cargo. My cabin had a little radiator very close to my berth, and a number of times, when I awoke, my bed clothes were frozen to the radiator, which meant I got a chill. By this time, on our way to St. John's, I was just about crippled. Even before I left Rose Blanche to go to

Halifax I could hardly walk. And being up to my waist in water all day with severe frost didn't help. Nevertheless, we arrived at St. John's with no damage, except for losing a few barrels of oil. I noted a protest on same after arriving. It was now around the tenth of March 1943.

While I was docking in St. John's, I noticed a man walking out on the wharf. He seemed very interested. When we were tied on, I was standing on the bridge and he looked up and said, "Are you Skipper Arch?" I said, "Yes, sir." And he said, "I guess you don't know me?" "No, sir," I said, "I don't believe I do."

Then he introduced himself. "I'm W. Wareham, of W.W. Wareham, Harbour Buffett. Can I talk to you for a minute when you're through with the Customs?" And I said, "Yes, sir. Come up."

After going through the Customs and talking with my owners a little while, Mr. Wareham came aboard, but I was in such misery that I didn't feel much like talking to anybody. We went in the cabin, and he made it clear that he knew this was my last trip on this boat, and he knew all about my past career and the vessels I had sailed. Then he said, "we just got the *Ronald George* bought in Lunenburg, and we were wondering if you'd be her master and go fishin' in her."

I said to myself, "Well, by jingo! This is a surprise!" The *Ronald George* was one of the best vessels up in Lunenburg, more powerful than any other schooner fishing in Newfoundland. I didn't jump right away. I just said, "Well, Mr. Wareham, I think I'll just think it over." "Oh," he said, "You won't have to think that over very long, will you? I'd like to know right away." And I said, "Yes. I think I'll make up my mind right now. Yes, I'll go." "Well, that's good," he said. "Now, whatever you want to do, tell me and I'll do it." And he explained that the vessel was still on a freighting trip in Nova Scotia, and it would be a couple of weeks before her arrival at Harbour Buffett, Placentia Bay.

It was March now, and all the other Newfoundland skippers already had their crews and were preparing for the year's fishery. So I said, "Well, the first thing you'll do is put it over the air that you have the *Ronald George* and I'm going as her skipper, and any fisherman interested, wanting a chance in her, should wire you in St. John's or wire me in Grand Bank." At that time every home around Fortune Bay, and perhaps Newfoundland, listened to the Gerald S. Doyle News Bulletin that came on twice each day with local news and messages. That was important to fishermen. It was the only way families had to contact their men out at sea. If someone was sick or died, they would send it through this local news bulletin.

I was still crippled from the chill I got on the boat during the winter, and I doubted very much that I could get away without seeing a doctor. The freighting boat I was on for a few months before Christmas happened to be in St. John's on her way to Grand Bank, and the captain was glad to give me a passage home. On our way to Grand Bank I had to have help to get my boots on. I hurried up to see Doctor Burke after I arrived home, and he said I had a bad chill in my legs. I was worrying now, and very much afraid that I'd not be able to go in my new vessel.

There was a bulk of telegrams waiting for me at home, all from men wanting a chance, and my wife did not know what to make of it all. She knew nothing about me taking this vessel until I arrived. Perhaps twenty were from men who had sailed with me the year before in the *Pan American*, and the year before that in the *Florence*, and in other years. I was glad to get their telegrams, because I knew the men well. Within forty-eight hours I had a full crew, and I wired the men I accepted and sent one to the others saying, "Sorry, filled up this time." And I had to wire Wareham's to put out another message over the air to stop the telegrams from coming in.

Fortunately for me, the vessel was delayed longer than expected. Had she been ready a week after I arrived home, I doubt very much if I would have been able to take her. She called after me in the first week in April, and I went around

216　　　　　　　　　　　　　　　　　　Voyage to the Grand Banks

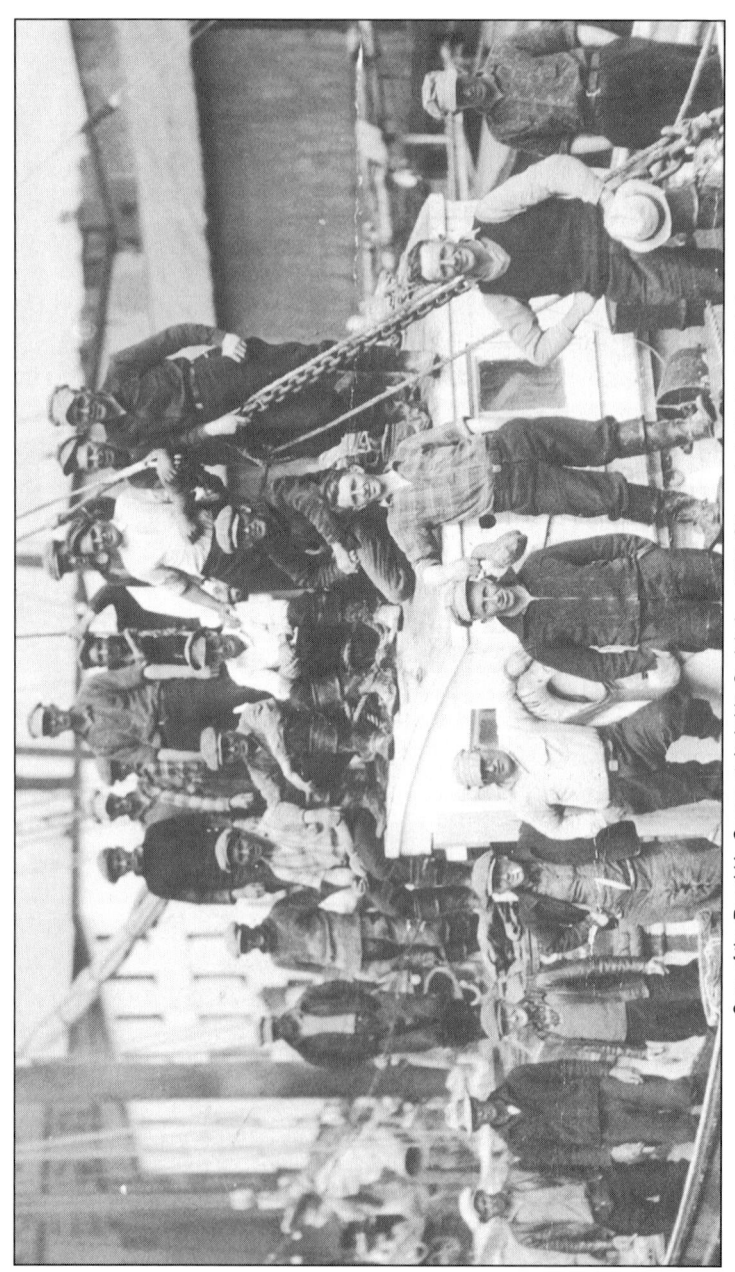

Crew of the *Ronald L. George* at dockside, St. John's, c.1944. Skipper Arch, third from left, below.

Fortune Bay to collect my crew and sailed for Harbour Buffett, where we prepared for the salt fishery.

I could say without contradiction that the *Ronald George* was then the best deep sea fishing vessel in Newfoundland. She was a wonder for her time. I'd known the vessel ever since she was built. She was fitted with a 100 horsepower Fairbanks Morse heavy duty engine,[1] and I felt sure I was sailing from one of the best companies in Newfoundland.

It was getting up towards May month when we sailed from Harbour Buffett for our first trip. The rest of the vessels had been fishing since the tenth of March, and quite a few had already landed a lot of fish. We only had time to fish two baitings before it was time to discharge and get ready for our caplin trip, but we were fortunate to find a lot of fish. With a good crew of men, we landed a large spring trip in the little time we had to fish.

The owners ordered me to proceed to Harbour Grace and sell our fish to Crosbie and Company of St. John's. We discharged the fish there and immediately left for Harbour Buffett, where the men took on board supplies for their families. I was dropped off at Grand Bank, and I gave my crew a few days for themselves. The mate, Waldram West, took the vessel to Bay L'Argent, and tied her up there while the men were home. Three days later, the vessel arrived at Grand Bank to get me, and we secured a baiting of caplin close by. We left immediately for the Grand Banks. On our first day fishing there we secured 200 quintals, and we got a large trip without weighing anchor. With this baiting fished, we proceeded to Harbour Grace again and discharged our catch. We fished two more baitings (caplin and mackerel), and secured another large trip which we landed at Harbour Grace. We fished our last trip, another good one, on the Labrador and landed again at Harbour Grace.

Despite getting started late on our spring trip, we ended up with only 500 quintals less than the highliner of the whole deep sea fishing fleet, and our owners and all the crew and myself were a pretty happy bunch. Everyone worked hard, but I never kept any man without sleep long enough to prevent him from doing his work.

You could see where every cent you made went

Another thing I can say without any contradiction, when we arrived at Harbour Buffett for settlement that fall, our owners gave us the very first statement my men and I had ever seen while fishing in Newfoundland. It was exactly like what we had while fishing in Lunenburg. Whenever we fished in Newfoundland before this, when you finished in the fall you never got a detailed statement about how much fish you had, or how much your bait and food bills came to, or anything like that. All you could get when you settled up was your own account. You'd have everything on credit in the summertime, and if you had any money coming, you would get it.

Wareham's gave us a statement from the first day we sailed until we ended in the fall, with every pound of bacon and every bit of ice, and any other expense we had to pay for, and pinned it up in the cabin where every man could see for himself what each man shared, what percentage the skipper was getting, and what orders were taken out. You could see where every cent you made went. That might not seem like much to some people, but can you imagine working hard all year and having someone else control your half of the catch and not give you any statement at all? Not a thing? It wasn't a very good policy. We got no account of how much bad fish there was or how much good fish. There was just a lump sum. Down at Grand Bank, you'd hear the men talking about that all summer long. But if they went up and asked for a statement, they didn't know if they were going to lose their jobs or not.

You'd be in your home port for about one week, back from your last trip in the fall, washing out your fish and then the schooner and dories, and putting sails and all your gear ashore. And you'd be sitting around the cabin or in the focs'l at night, nothing else to do but chew the fat and wonder how much we were going to make this year. You didn't know right up to the last day. If it was late in the day, and you'd scrubbed your dories, the last thing you did, and the next day you were going to settle up, well, something was bound to happen among the men in the focs'l. Maybe you had heard how much another schooner's crew had made for so many quintals. When the time came for you to settle up, perhaps you had more fish and expected a little more, but didn't make quite so much money. Well, that was hard to understand. There wasn't anyone to tell you anything, and then one fellow would say, "I'll go right up to the office tomorrow and find out something."

Well, he went up and got no satisfaction, and probably never got back at that same schooner next year. Because the owner probably told the skipper, "Look, boy, you won't have that fellow next year. Don't have him. He's too busy, that fellow." That happened several times. And I know one skipper who went up to his owner about the food, but I don't think they had a long conversation.

I don't know that they also did this with their smaller vessels, but I think Warehams got the idea of using settlement statements during that first week I was in Harbour Buffett fitting out the *Ronald George*. We used to talk things over up in their office at night, and they asked me a lot of questions, as I did them. They wanted to know all about how things were done when I fished out of Lunenburg, and I mentioned the statements we used to get there.

In Lunenburg, when you came in off a trip, regardless of whether or not you had a single share in the vessel, the major owner would come along and talk to you just as if you owned it all. I had only two shares in the first schooner I commanded, and the owner would ask me what I was going to do. It seemed so strange to me. I told them how it was up in Lunenburg. I'm not saying that's why we got it, but you didn't have to ask them much, you didn't have to coax them to do something.

We were at Harbour Buffett for a few days putting our

fishing outfit ashore and cleaning the vessel, etc. Then the whole crew took their supplies from our owners. It was almost a full load by the time it was all on board, and we sailed for our homes in Fortune Bay.

After taking my men to their homes, I went to Grand Bank for a few days in the vessel. There I took on a coasting crew, went to North Sydney for a load of coal, and discharged it in places in Placentia Bay, such as Clattice Harbour, North Harbour, Baine Harbour and Harbour Buffett. When that was done, we sailed to Prince Edward Island for a load of produce, discharged it at Harbour Buffett, then tied the vessel up there for the winter, and arrived home just before Christmas.

The Year 1944. The early part of January I received correspondence from my owners at Harbour Buffett, asking for my opinion about going fresh fishing this year instead of salt fishing. I was delighted to get a crack at it, because I never did much fresh fishing. They said, owing to the war, that there was a large demand for fresh cod fillets in Britain, and Job Brothers at St. John's would pay a satisfactory price for it and take all we could catch. I wrote them and said it would be satisfactory to me, and, in the meantime, I think we should get an early start on, because quite a few changes had to be made on the vessel, such as new fish bins in the hold and on deck.

The last week in February a boat called for me and five other men. I went for the vessel, came back to Fortune Bay for the rest of my crew, went back to Harbour Buffett again, and made all the necessary arrangements for fresh fishing. Some of my men were with me last year, but I had changed quite a few — after coming to mutual agreements, including my cook.

Around the middle of March we started fishing on Rose Blanche Bank, and sold our fish to Job Brothers' fish plant at Port aux Basques. The last week in March our owners advised me that Job Brothers would be ready to take our fish at

St. John's as of the first of April. We did well since we came up on this coast, discharging our catch every two or three days. The first week in April we took on a supply of ice and bait and sailed for the Grand Banks. We were very fortunate during our first two days fishing and loaded our vessel in only six days from the time we left Port aux Basques until we arrived at St. John's.

There was quite a bit of excitement on the south side of the harbour at Job Brothers' fish plant. This company had been buying fresh fish from the shore fishermen for a few years, but we were the first deep sea fishing vessel to land fresh fish in St. John's harbour that was caught by dories on the Grand Banks, and many business concerns were watching this new adventure with keen interest. Our owners from Harbour Buffett, and all the top brass from Job Brothers and Company, were there to meet us shortly after learning of our arrival.

None of the men had ever been fresh fishing before

It was quite a change for us too, going from salt fishing, which we had been doing all our lives, to fresh fishing. None of my men had ever been fresh fishing before. It had been going on in Nova Scotia for a number of years — deep sea fresh fishing — but not in Newfoundland. Discharging would begin immediately upon our arrival in the morning, and we'd be on our way for the banks again by 12:00 or 1:00 o'clock the next day with a supply of bait and ice. The men worked hard discharging the fish, but they didn't work on it at night, and most of them took the opportunity to have a few hours ashore for a little recreation. After all, we were all in the war and one never knew when he sailed out through the Narrows of St. John's harbour if he would ever return. The war was still on and there was much activity and many large and small warships in the harbour.

Our arrival from the banks, with our big trip of fresh fish, to find our owners and other people waiting there to meet us, reminded me of another day: Our return from the banks while fishing out of Grand Bank in the *James and Stanley*, with all of our gear, cable and anchor lost, and our sails torn to pieces, and no fish to go towards it. Quite a contrast, but life is made up of failures and successes.

The following night after leaving St. John's, we were at our destination, began baiting our gear, and had it all set in the water by daylight. We were fishing with twelve dories, two men in each dory, and were twenty-eight in crew, including the cook, deckhand, engineer and myself. Shortly after daylight and after the men got a snack, I dropped them on their gear again, and by the time I dropped my last dory I noticed the first dories had signals up, which meant we were on a lot of fish again and they had their dories loaded, calling me to them to be discharged on the mothership. It was a lovely, fine day, each dory had sixty lines out, and, while I was steering back to the first dory, I could see they were all getting a lot of fish.

We were all busy that day, including the cook, deckhand, engineer and the skipper. By the time we had our gear all back that evening, the deck was so full of these large, Grand Bank fish that we did not have room to hoist in our dories, so we anchored and tied them all on the stern of the vessel. There's quite a big difference between dressing fish for salting, and for fresh fish. On fresh fishing, all you do is take the gut out, leave the head on, and then ice it in the hold of the vessel. After the fish were all gutted and iced, the men went for their rest, then went through the same routine again, such as baiting their gear, etc. The next day, there was just as much fish as on the first. We had another load and left for St. John's around four o'clock in the afternoon, and arrived just in time to get through the wartime gate at the Narrows before dark. We were loaded deep, and what a pretty sight steaming up through St. John's harbour. We were just five days from the time we left port until we arrived again, and weren't tied up long before the boys had gone ashore for a little recreation.

On our way in from each trip on the banks I would go around and ask every man what he wanted, and I'd put it down in my book. After we tied up by Job Brothers' plant on the south side, I'd go over to W.W. Wareham's office. If a man wanted $10 or $100 sent home to his family, whether the total came to $1000 or $3000, Mr. Wareham would write out the cheque for me and I would go to the bank and change it for cash. Back aboard the schooner, all the men gathered in the cabin, and I'd take out my book and give each one what he asked for. I kept the books, and when we got back to Harbour Buffett in the fall we'd go and settle up. Harry Wareham, the youngest brother, was in the office, and he'd get on one side of the desk and I on the other. Now, if I was five dollars out on the thousands that I handled, I'd have to pay. But I never was.

The rest of April and May months continued to be very successful and many bumper trips were brought in. The caplin were in by the first week in June and we carried on the same routine as in other years when we were salt fishing. We went to Harbour Buffett, our home port, where the men took on supplies for their families. Without losing any time, we were on our way to Fortune Bay, where all my men belonged, and had a weekend home and an extra day with our families, then took on board a small amount of caplin, compared to what we would take for a baiting when salt fishing. The luck we had all spring continued after arriving on the banks with our first caplin. We fished only one-and-one-half days and took enough to get in St. John's and out again before the weekend.

My crew and I believed in making your hay when the sun shines

We gave Job Brothers and our owners plenty of surprises by such quick trips. All of my crew and I believed in making your hay when the sun shines. Our company had three more smaller vessels landing fresh fish, which meant quite a bit of activity at Job Brothers' fish plant on the south side of St. John's.

Our bumper trips continued through the season, and we ended our voyage in the first week in October. By then the weather was getting too stormy to fish in dories on the Grand Banks. Our year's work ended with the largest amount of fish ever caught in dories and brought to a port in Newfoundland in one season, and the most money ever made by a vessel's crew in one season. Job Brothers had already hinted that they were thinking of going big into the fresh fish business and this may lead them to get into dragging.[2] And they said, "so we're going to be after you to go skipper of one of the trawlers."

After all of my men were carried around to their homes, we took on a coasting crew and went freighting again. My cook changed over to mate. We went to North Sydney for a load of coal, and then to Prince Edward Island for a load of produce. Our owners decided to put a larger engine in the vessel, and replaced our 100 horsepower Fairbanks Morse engine with a new, 200 Atlas Imperial, and now we could clip along at eleven miles per hour.

It was getting along in December when we discharged our cargo of produce at Harbour Buffett, and took the vessel to St. John's to have the new engine installed at the C.N. (Canadian National) Dock. When the vessel was tied up for the winter at the dock, the crew went home, and I arrived home about a week before Christmas.

During my first spring as master of the *Ronald George*, I bought a new house about three weeks before the vessel called for me, so I had a lot of work to do before I started fishing again in the spring. I'd had one of the best crews of men that could be carried on a fishing vessel this past season, and they were all going back with me again next season. And our company had confirmed that the *Ronald George* would continue fresh fishing with Job Brothers.

As I mentioned, we had had a good, successful season this past year, and all our men and I were delighted in the change from salt fishing to fresh fishing. We also had high praise for our company, W.W. Wareham and Sons. There were three brothers and the father, and I don't think one

A Wonder for Her Time

Ronald L. George at wharf in St. John's, c.1946. She was salt fishing, had just been reprovisioned, and was ready to sail.

could meet better, more honest men for their business on this earth.

Little happened to us this year, with the exception of a few narrow escapes owing to the war, and a couple of near collisions with large icebergs while steaming home with a load of fish in dense fog. And large ship convoys steaming over the Grand Banks forced us to get out in our dories several times, mostly by night in dense fog, during the last two years.

There was a convoy coming right towards us

Once we were anchored with a strong wind and a big sea running. The watch called me at about two o'clock in the morning, and said there was a convoy coming right towards us. They could hear the ship's sirens. It was dense fog. All of our men were called out of their berths and dressed in their oil clothes, ready to get out in our dories at any minute. They came so close that I ordered all our men to leave for fear of being cut down. One of these large ships would just have to touch a small ship like ours and she would go to the bottom very quickly. Quite a few fishing vessels had already been run down by convoys since the war started. Only a short time before, another convoy sank a large, Nova Scotia fishing vessel and a number of the crew went to their deaths. Only the engineer and I remained on board. The engineer kept the air coming for the siren that I was sounding all the time. It was a very bad night to have to call men out of their comfortable berths and into an open dory in such seas as were rolling that night.

It seemed to us like one of the largest convoys in the war. The ships came so close on each side of us that we could see their glimmer at times. Finally, after what seemed an eternity, one monster came so close, it seemed as if I could put out my hand and touch the ship. I looked up in the sky to holler out and ask, "How many more to come?" A voice came back and said, "We are the last ship, Captain." I knew then that the engineer and I were safe, but what about the twenty-six men out in the dories? They were all spread about like chickens around their mother. I gave the signal, four long blasts, which meant come on board, and they began coming, one after the other. They were out in the dories for almost three hours. Six dories were out. Two had five men each, the others four men. I was a happy man when the last dory arrived. After coming on board, the men explained that several times they could touch the ships with their oars. The war years were a dangerous six years for fishermen on the banks and many lives and vessels were lost to convoys, but where was it, and what was it, that held no danger?

The first week in February 1945, I took four men and went to St. John's — my mate was already there — to bring the vessel to Harbour Buffett, as the installation of the engine was about completed. February 26th, we left St. John's with a brand new, 235 horsepower engine in our vessel, *Ronald George* — the wonder vessel. We not only had the best bank fishing vessel in Newfoundland now, but the most powerful fishing vessel in all of Newfoundland.

In the early part of the season, we were fortunate enough to have the first ship-to-ship and ship-to-shore radio telephone ever put on a Newfoundland-owned banks schooner fishing out of Newfoundland. One other vessel had this equipment, but it was in the vessel, already installed in Nova Scotia, when it was purchased. And that vessel was then fishing out of Nova Scotia. Some Nova Scotia vessels had had the ship-to-ship radio for quite a few years now, but this necessity and convenience was long overdue for Newfoundland fishermen, and words cannot describe its great value. Our other vessels had receiving sets, but no transmitters. By the end of this year's voyage, radio telephones were also ordered for the other vessels for the next season.

The radio was great for talking to one another out there. 'Twas like another life. Before we had a transmitter, on stormy days, you'd just be sitting around, looking at one another listening to the radio.[3] But with the ship-to-ship radio, now you could talk to one another all day long. It was company and a great thing if you had to make a distress call. Compare it to fishing on the Grand Banks in schooners, when a dory went astray with two men and they might get aboard another one. Without the radio, you would lose your trip over it, because you had to come to land and report it after a certain time. With radio transmitters, before you knew it, perhaps the skipper of the other schooner would be calling you, "Arch, your two men is aboard, boy."

It didn't do so much for fishing, because very few tell the truth out there anyhow. When everyone had radios, you couldn't depend on what another skipper would say. If he said, "No, boy, there's nothing at all here. 'Tis not worth your while to come here.", he might just as well have taken 200 quintals that day. And there were days when they'd say there was plenty of fish where they were, but perhaps there was none. That wasn't done too often, but it happened.

Some skippers had their tricks to throw you off, to get fish

There was no way to tell lies without a radio. With the banks schooners, if you were on the Grand Banks and only two miles from a man, the only way you could tell if he was getting fish or not was when the weather is clear. Then you could see his lights at night, that he was dressing fish, working all night. If it was thick fog, and it most always is on the Grand Banks, you wouldn't know what the other fellow was doing if you didn't cross a man close enough to speak to him in the vessel. And then some skippers had their tricks to throw you off, to get fish.

Once, Skipper Mawg (Morgan) Matthews was fishing at anchor on the Grand Banks. It was toward the end of the caplin trip, about when the squid struck the Grand Banks and the whole crew would jig their own bait. Fresh squid would get two or three times as much fish as the last of your iced caplin. Mawg was there getting a lot of fish on his last caplin when this other schooner came along to speak to him. When he saw it coming, he placed some men on the stern quarter and had them pretend to work their squid jiggers, to make the other skipper believe he was getting squid. But there wasn't a squid in the water, and I don't think they came in at all that year. Mawg told him there was a lot of fish there, but he didn't say he was getting it with caplin. So this fellow anchored a berth away from him and put his crew to work with their jiggers too. And he never set his gear that day with the caplin he had, because he thought the fish weren't taking it. By dark, he had no squid and hadn't set his gear either. And there was Mawg catching away with caplin. I'm not an angel, but my conscience wouldn't let me do that. You didn't have to do things like that to make a living.

Another time when I was in the dory with a skipper I won't name, we were fishing about 120 miles off on the banks on a small patch of shoal ground called the 'Hump'. There were always a lot of fish there. It was Sunday night and we'd already taken about 200 quintals while anchored there on Saturday. But another Grand Bank schooner was anchored nearby and our skipper thought it was taking more fish. He called us out to bait up that Sunday night, and first had us heave up the anchor, to jog around this other schooner until daylight, so he'd drop us off in just the right spot while the other men would still be heaving up theirs. We'd head them off and be on that spot of fishing ground. Normally, you'd bait up, wait for light, and then heave up to change your position.

But the other skipper was too smart for him. He saw us joggin' around, dodging back and forth, baiting up our gear at the same time so we'd be ready to set our gear before him. At daylight he was still there, anchored we thought. But this man was cute enough that he told his men to heave the anchor just so many fathoms from the bottom. So she was drifting off while we were joggin' around her. At daylight, we clapped our eleven dories out and set our gear. Then our skipper asked, "How come he be so far away from us as this now?" We didn't bother to sound before. And when we went out and took in our gear, there wasn't a cod fish on it, just some black dogfish and other, old, queer fish. Then we sounded and found 300 fathoms of water, that we had driven out over the banks.

Meanwhile, the other skipper anchored up there, fished, and had the biggest kind of day. I don't think our skipper ever forgot that, ever forgave him for that, because it made him a laughing stock. It was a mean thing to do. I couldn't do it, and I know many other men wouldn't either. There was no reason for us to heave up that night after we had taken the bait off our gear. But our skipper said we'd be there to set our gear first, just in the right place when the other skipper was only heaving up his.

We talked about a hard struggle up to 1945, and all of my family and I were proud to have such an outfit, the *Ronald George*, to make a living.

It was late in the evening when we left St. John's for Harbour Buffett, with a load of freight for our owners. Frederick, we always called him "Freddie," one of the brothers in our company, was with us. A very nice man, and the entire crew had admired him ever since we joined W.W. Wareham and Sons. The *Ronald George* made excellent time with all her sails and the new engine. The following evening, as we rounded Cape St. Mary's, a southwest storm caught up with us, and there was thick snow. We arrived at Harbour Buffett by four p.m. on the 27th of February, my birthday.

One of our crew was another of the company's captains, Captain Jacob Thornhill from Grand Bank. He joined the firm during the second year I was with them, and sailed the *Governor Anderson*, a vessel smaller than the *Ronald George*. He sailed her one year, and now the company had bought the *Robert Esdale*, another vessel the size of the *Ronald George*. Captain Jake, as he is called, would take her salt fishing.

I was given a big surprise birthday party at Freddie's home on the night we arrived. All the company's brothers, and other people around, joined in on the happy occasion. There was a wind and snow storm that night, which made the party more enjoyable. There was still no radar then, so we might easily have been forced to hove out in Placentia Bay.

By the time we had our freight discharged and other necessary work finished, it was time to go to Fortune Bay to gather up my crew for fishing again. I had the same crew as last season, except for a couple of men, and we were to land our fish at Job Brothers again, selling it fresh. Around the third of March, a Saturday afternoon, we left St. John's and reached Grand Bank in the night. We were leaving Monday morning to go in Fortune Bay for our men.

While the vessel was tied up by the pier in Grand Bank Harbour for the weekend, the *Clarenville*, one of the C.N.R. boats, came in and collided with our vessel, cutting in through the stern as far as the cabin gangway. Anyone coming out of the cabin at the time would have been killed instantly. I was notified and immediately went on board, and what a sight to see, after just having our new engine installed! I notified our owners. The father of the brothers was living in St. John's, so it took little time to contact him, although there was only one phone for communication at that time. He authorized me to take charge and get the damages repaired to my satisfaction.

Monday morning, I engaged a Mr. Grandy, a contractor, at Grand Bank, and he and his sons went to work immediately. We were delayed ten days, and had to put a new stem on the vessel from the water's edge up. The C.N.R. was responsible for all damages, including the estimated amount of fish we would have caught during the ten days and for the wages of the six crew members on board while the repairs were made. It cost quite a bit by the time they were completed, but Mr. Grandy and his sons made her as good as new and received many compliments for their fine work.

We lost no time getting ready for fishing after arriving in Harbour Buffett with our men. The other vessels were all on the banks by then, and it looked as if the war would soon be over, which would be a good thing.

Our first trip to the Grand Banks that season was very

successful and we loaded in two days. The weather was excellent, and we used almost twice the amount of gear we normally did. We continued to bring in bumper trips all that spring, and in one week in particular, we broke all of our previous records.

The Lieutenant Governor, Campbell MacPherson,[4] made a trip with us on the Grand Banks that year and, one night, as we neared our destination, before he went to his berth he made me promise to call him when our men began working. It was a fine night, and I could see the next day was going to be good as well. There was some fog, but if it's calm, no matter how foggy it is, we call it fine. Two o'clock in the morning, I called him and we began dropping our dories. Every dory crew had their kerosene torch with them. By the time we dropped one dory and the other was ready to be dropped, we were out of sight of the last one. After the twelfth and last dory was dropped he came to me and said, "Skipper, do you think you will ever see them again?" By the time it began to get daylight I was picking up all the dories again. He got a big kick out of it when I'd say, "They'll have their dinner, now, when they get on board, while we are steaming to the first ends they dropped." We had our load, landed the Lieutenant Governor and discharged, and went a short distance off on the bank and secured another trip with enough time left to discharge within the week, which gave us two trips in one week.

On another occasion, we left St. John's one evening and steamed 200 miles off on the Grand Banks, twenty hours steaming distance. It was a Thursday morning when we sighted a vessel and went to her. Two men and I went on board to get all the news. It was the *L.A. Dunton* from Grand Bank, and her master was Captain Alex Smith, the biggest fish killer in Grand Bank. He was salt fishing, and told me that there was quite a lot of fish in this vicinity, and he would be leaving tomorrow for Grand Bank.

It was a beautiful evening, and we had enough time to set our gear before dark. But the next day was going to be so

Dories returning to schooner *Ronald L. George* towards evening on the Grand Banks. Their gear has been set and will be left overnight. c.1945.

good and we were fresh fishing, so I decided to set our gear at the usual time — 2 a.m. — on Friday, because the faster one gets the fish off the trawls, and the less time it lays on the bottom, the better the product. And, by Saturday, our vessel was filled again! Monday afternoon, Captain Alex Smith was going through Fortune Bay at two o'clock in the afternoon, the hour when Gerald S. Doyle's (firm) always gave local news bulletins, and the first thing that came over the air was, "The *Ronald George* arrived St. John's, Sunday evening, with 300,000 pounds of fresh fish from the Grand Banks!" Capt. Smith said to me many times since that both he and his crew could hardly believe it, as it was only Thursday evening when I was on board him, and we only began fishing on Friday morning.

It isn't hard to understand how important it was to switch from the dummy schooner to the power schooner, and how hard it was for the last man to struggle along those last few years with no power at all.

We were all glad the war was over, but sure sorry to give up fresh fishing

We kept landing bumper trips that spring, but within a month after the war ended the bottom fell out of our fresh fish market and we had to switch to salt fishing again. We were all glad the war was over, but sure sorry to give up fresh fishing.

In no time salt was on board and we were on the Grand Banks again catching just as much fish as ever, only it was quite different now. The fish was caught, split and salted, so we'd be on the fishing ground much longer. We fished for six days, had another good trip, and left for Harbour Grace, to sell our fish to Crosbie and Company. After arriving at Harbour Grace, I decided to take only half of our catch out, to give us sufficient room to land a larger catch on the next trip. Our voyage ended in the first week in October after landing three large trips of salt fish at Harbour Grace. Our men had another big year, the third in succession.

After having our last trip discharged, we went to Harbour Buffett as usual, where the crew got their final settlement, and I landed them at their homes by the 20th of October. We then took on a freighting crew and went freighting during the rest of the fall. My cook changed his position from cook to mate as he did last year. It's quite a change for him, after cooking for twenty-eight men during the entire fishing voyage. He was a number-one seaman and had been mate a number of years in the foreign trade. And he had foreign navigation. His home was at Grand Bank and our two houses were close to each other.

A few days before Christmas we arrived home after landing loads of coal and flour from Sydney, and a load of produce from Prince Edward Island. All the freight was landed at Harbour Buffett. W.W. Wareham's was the largest business company in Placentia Bay, and they had a number of smaller, four to six dory vessels, and the *Ronald George* and *Robert Esdale*. The *Robert Esdale* also landed a good voyage this year.

In 1946 we began fishing around the usual time, taking on our men in the first week in March. We were salt fishing that year and landed our catches at Harbour Grace, and had another big voyage, a total of 7000 quintals. After finishing the voyage and landing the crew, we went freighting again that fall, and my cook changed his position as mate once again.

Our company advised me we would make three freighting trips, the same as last fall. Like the year before, our last trip was to Prince Edward Island for a load of produce, and we arrived at our homes at Grand Bank about ten days before Christmas, after mooring the vessel at Harbour Buffett for the winter.

The next year we arrived on the grounds to begin fishing again around the middle of March. I had the same crew of men with me, including the cook and engineer. My engineer had been with me ever since I took charge of the *Ronald George* and we got along fine together. By the time our spring trip was over we had landed 2500 quintals at Crosbie's and

Aboard the *Ronald L. George*, 1945. On left Charlie Parsons, cook.
On right, John Barnes, crewmember.

Company, Harbour Grace, and were amongst the top highliners again.

We landed a large caplin trip at Harbour Grace, and proceeded to North Sydney for a baiting of frozen mackerel and went direct to the Grand Banks. 200 quintals were secured on our first day of fishing, a fine start on our trip. After our bait was fished, we went to Harbour Grace to take out some of our catch in order to lighten our vessel for the remainder of the trip. Then we proceeded to North Sydney again, took on another frozen baiting of mackerel, and went back to the Grand Banks, in the same position we fished on our last baiting. Back in Harbour Grace again, our catch on these two mackerel baitings weighed out to 1800 quintals.

It was only the 20th of September, so there was sufficient time left for another good trip. After quickly securing another baiting, we went to the Grand Banks again and were fortunate to have all good weather, and secured another successful baiting. It was near the first of October, and as a rule the weather is bad by this time of the year for dory fishing, but this fall it was good. I remembered one year when I was fishing with my cousin, Captain Reuben Thornhill, and we took on board a baiting of squid and sailed for the Grand Banks on the first day of October and caught 800 quintals, and arrived Grand Bank on the 15th. I thought that could probably be repeated, so we went directly to St. John's for another baiting and sailed to Cape St. Mary's.

For the next ten days the weather couldn't have been better in the summer. We were the only vessel fishing there at the time, and the fish were abundant. In six days our salt was used up, and, after taking back our gear, we left for Harbour Buffett with another big day's work of fish. At Harbour Buffett we discharged 300 quintals and took on board sufficient salt for another few days of fishing. The weather continued to be good for another week, and by this time our vessel was filled again and we sailed for Harbour Buffett, where we put the fish we discharged a few days before on the deck, and sailed immediately for Harbour Grace.

The *Ronald George* steaming for Harbour Grace with a full load and fish on deck made an unforgettable picture. But the following morning, steaming into Harbour Grace, I was too close to the buoy on our starboard side and our vessel was drawing quite a bit of water, so she went aground. Most of the fish on deck had to be jettisoned to lighten our vessel and, just before dark, we floated off and tied up at Harbour Grace. Once again, we had landed over 7000 quintals for our season's work.[5]

A better company couldn't be found . . .

During the year and a half we landed fresh fish at Job Brothers, St. John's, we had five of the largest voyages in succession that ever could be expected from dory fishing. It was quite a few million pounds, and our company told me many times that they never, ever regretted purchasing this vessel from Lunenburg, nor did they ever regret putting the new engine in her after the first two years with the old one. I had no regrets either, and I can say without any contradictions at all that a better company couldn't be found from one end of our island to the other. And another thing I can say with a clear conscience, during the five years I sailed this vessel, I never ever heard one man lay a complaint against W.W. Wareham and Sons at Harbour Buffett.

After discharging our fish we proceeded to Harbour Buffett and went through the same routine as in other years. While there, a meeting was called at our company's office with the three brothers — Freddie, Leman, and Henry, the father, W.W. Wareham, and myself. It was decided that I would land my crew in Fortune Bay, then take the vessel to Grand Bank and engage Grandy and Sons, the same contractor who repaired our vessel when the *Clarenville* collided with her at Grand Bank harbour a couple of years ago, and have a new top, including decking and rails, put on the vessel from the water's edge up. Because she was getting on in years. But first I would go to St. Pierre and dock the vessel for a good inspection of her bottom. After landing my men, these plans were carried out, and the vessel was brought to Grand Bank, where Mr. Grandy and his sons took on the big job. I was to supervise the work and have everything to my satisfaction.

Chapter Eleven

Millions of Fish

In this present chapter Captain Thornhill recalls his first four years — 1948 through 1951 — as master of the *Fearless* — later renamed *Blue Foam*, a 140 foot steel hulled groundfish side trawler or 'dragger.' He reluctantly left the *Ronald George* — "the pride of Newfoundland banks schooners," and its Harbour Buffett owners, to become a dragger captain for Job's Fisheries Ltd. of St. John's. Soon the *Fearless* touched in at Grand Bank, where Arch loaded his wife, children and furnishings aboard, and sailed them to a new home at St.John's.

Arch's story is largely silent about major changes under way in Newfoundland society at this time, among them a growing cash economy, rising expectations and materialism, Confederation with Canada, and growing government intervention. Government was preoccupied with industrializing Newfoundland's economy. As Newfoundland fishing firms entered fresh-fish markets, they introduced groundfish trawlers, cutting lines, refrigeration, and created year-round fish plant wage labour opportunities and needs at trawler ports.

For government, fish plant and trawler ports became "growth centres" and the focus of their community resettlement schemes away from the inshore/nearshore fishing for the salt fish trade. That sector was believed anachronistic. Despite mechanization of its vessels it remained seasonal. Hundreds of its small outport communities, like Anderson's

Cove, Arch's childhood home, were distant from modern fish plants. They 'rubbed along' with salt fish production until their residents either gained access to plants or moved away and abandoned them.

Diesel powered groundfish trawlers had revolutionary implications for Newfoundland's offshore fishing, fishermen lives, and their communities. Work regimes, roles, skill and marine ecological knowledge requirements, incomes, and personal lives ashore were changed. There was a quantum leap in fishing range, effectiveness, catch scale and reliability, and two way radio transmitters enabled close coordination of their target species, and catches and landings with plant requirements. Arch's story gives us glimpses of an emerging "vertically integrated" system of trawler fleets and plants — both owned by non-fishermen interests ashore — that became the new standard.

The ecological range of their fishery expanded. Groundfish trawlers enabled Newfoundland's industry to move beyond its historical narrow focus on cod to exploit other target species, like haddock, redfish, and flounder. Trawler skippers had to learn to fish new species and grounds, at depths they had not fished before. And the competition for fish on offshore banks increased. Canadian, and foreign trawler numbers all grew, as did the intensity of their fishing. Pity the fish and their sea bottom home. This industrial fishing expansion trend peaked in the late 1960s with arrival of giant foreign factory-freezer trawlers that swept the grounds, taking everything, leaving nothing.

Arch describes how early groundfish trawling destroyed extraordinary quantities of undersize haddock. Government fishery science enters his account at this point for the first and only time. In response to Arch's concern about their future, an unidentified scientist tells him the ocean's fish stocks were inexhaustible, a view widely held by scientists then. This anecdote rings familiar today when fishermen rage about how their deep frontline knowledge is ignored or dismissed by fishery scientists who have yet to discover how to interpret and apply it.

By the late 1960s Arch had retired from trawling, and haddock were commercially extinct. Later, regretting the waste years before, he told a young provincial government fisheries official how shovels were specially supplied his trawler to speed discarding small fish overside.

International fisheries "management" bodies, like the International Commission for Northwest Atlantic Fisheries (ICNAF), were in their infancy at this time. Although established in 1948, ICNAF remained but an information collecting body for decades. The information was of doubtful reliability, and catching technology and effort offshore were unmanaged. Market demand, catching technology innovation, profits, and employment imperatives led fisheries science knowledge and stock management for some time.

Elements of the old banks schooner fishery persisted in the social organization of Newfoundland's early offshore trawling. Trawler fishermen remained unorganized labour, classified under the fiction of independent "co-adventurers." Their earnings were based on the share value of their landings. Yet the value of their landed catch was determined by trawler owners, and they were subject to the arbitrary authority of owners and captains.

Despite improved earnings and work conditions, the intensive work regime weighed heavily on the men and their families. Dragger fishing was pursued through all seasons of the year, trip after trip, with but a few hours turn-around time, and without scheduled holiday leave time. And few, if any, insurance or other benefits were provided despite the work's hazards. However, these choice fishing jobs were scarce, and men were in a poor position to press complaints. As one old deckhand put it, "You couldn't look sideways at the skipper. You might get a swat in the head and sent ashore." They had left the penury of the fish merchant credit system Arch described when speaking of first going to the

Grand Banks in 1918, over thirty years earlier, but equity in employment relations had changed little in the new industry at mid-century. Industrial change is like that; ragged, unsystematic, and often dogged by social practices vested in old power relationships.

Trawlers did improve work and living conditions at sea. Men were better accommodated and provisioned, and it was much safer than dory fishing from schooners. Engine power, and new navigational and communication technology all enabled reduced life risks at sea. But the work remained arduous and dangerous. Limbs and lives were wrecked by winches, writhing steel cables, faulty blocks, rough seas, slippery decks and winter icing. Accidents and loss of life were facts of life for all offshore fishermen. The consecration of their fish with blood, sweat and lives continued. And boat owners, now able to exercise authority via radio, changed the brotherhood of the sea equation. The skipper even lost freedom to act as good Samaritan toward other seafarers in crisis. He must seek his boat owner's permission first.

1949-1951

I was at Grand Bank for about a week when I received a phone call from the Superintendent of Job Brothers in St. John's. I hadn't seen any of them since we gave up fresh fishing after the war. First he congratulated me on landing all the big trips since leaving them two years before, then he said, "We haven't seen you since, but we know all about your success." I wondered, "Now, what's coming?" And he said, "Remember when we used to hint to you that some day we might have draggers? Well, we have them now. Two brand new steel trawlers; one has been used almost a year and the other six months." They registered 400 tons, had 800 horsepower engines, and were both at Boston. Then he asked me if I would be willing to go as master of one of them.

It would mean being away from home winter and summer

Now I was in a fix: I already had the best, most powerful vessel sailing out of Newfoundland at the time, and was employed by the finest company I had ever known. And there were other things to consider: It would mean being away from home winter and summer. He wanted my answer as soon as possible and within twenty-four hours I had made my decision to join Job's. Whether it was right or wrong remained to be seen. It was strange then to even hear the word "dragger" mentioned.

Our three youngest children were at home at the time, and they couldn't understand what it was all about. Our oldest daughter, Catherine, had gone to Mount Allison University at Sackville, New Brunswick, to study music, and our youngest daughter, Florence, was quite upset when I told her we would be fishing summer and winter, and probably get home only once or twice a year. But I promised her that, if Dad could make a success of this new venture, it wouldn't be too long before the family would move to St. John's and they would become "Townies."[1]

It was difficult to wire my present employers and give my resignation. No captain and crew ever had a better relationship with their owners than we did. A number of telegrams were exchanged, but people like the Warehams knew what was right in a case like this, so my resignation was accepted with good will.

Job's called me to St. John's a few days later and arrangements were made for me to take seven men to Boston. But first I went home again for a few days. I had already hired my previous crew from the *Ronald George* for the next year. But that meant twenty-seven men and myself on the vessel for dory fishing, and the dragger only needed nineteen. Fortunately, Baxter Blackwood, the captain of Job's other dragger, gladly hired the men I wouldn't need, and I recruited four with dragger experience.

I left Grand Bank on about the 20th of November with a skeleton crew, including the same cook I had had for the past

Skipper Arch and merchant, Freddie Wareham, at Harbour Buffett, c. 1943. The *Ronald L. George* had put in for supplies.

four years. We joined the train at North Sydney and stopped off for the weekend at Sackville, to visit my daughter, and we were both happy to have a little time together. A few days later we reached Boston.

I had never been on a trawler in my life

I was really excited to see my new ship. It was quite a difference from the schooner. The method of fishing involved in dragging was also new to me. It was quite unlike dory fishing. And I had never been on a trawler in my life. I made this quite clear to my company before accepting the responsibility, and they seemed to have enough faith in me to take the chance.

After being at Boston for a week, I decided to take a trip with the top dragging fisherman in the States, Captain Richard (Dick) Dobbin, a Newfoundland man from Mt. Carmel, St. Mary's Bay.[2] He had been living in Boston for a few years, and his trawler's name was the *Flying Cloud*. Although we were on the banks for a week and caught a fine trip of fish, I discovered you couldn't learn very much in only a week. One important thing I did learn was about how strong the wind would be when it was time to stop fishing and take your gear on board.

The Queen Mary couldn't have looked better

While I was on this trip my men were busy putting everything including the engines on our trawler, *Fearless*, in first-class condition, and by the time I returned the *Queen Mary* couldn't have looked better. Our boat had the best of sleeping quarters; both the captain's and owner's staterooms had a wash basin and shower. The furniture was all mahogany. On the top deck there was a large wheelhouse with the captain's room attached, and another room with three berths and a settee, and a mahogany table in the middle of the room. This was first called the "owner's room," but was later known as the "guest room."

There was quite a bit of brass in the wheelhouse, and the boys took great pride in shining it. The large galley had a steel sink for the cook, and a large, walk-in deepfreeze, and there was a steel grating to walk around the engine room on the first deck. The *Fearless* was 140 feet long and, besides its 800 horsepower engine, it had two eighty horsepower auxiliary engines used for electricity and driving the main winches. The ship also had all the necessary navigational instruments available at the time.[3]

This may not seem very interesting these days — twenty-eight years later — but it was at that time because this boat and her sister ship, the *Challenge*, were the first and only ones of their size and type in Newfoundland.[4]

This was something new in Newfoundland

We left Boston for St. John's around the middle of December, and enjoyed a pleasant passage home. The *Challenge*, under Captain Blackwood, left four days earlier and also had a pleasant trip. The St. John's news media were busy for quite a while after the two trawlers arrived, because this was something new in Newfoundland. By this time there were two other fresh fish plants in Newfoundland, but they operated on a smaller scale. One plant on the southwest coast only had a small, inshore dragger, catching mainly redfish.[5] The other was at Burin, which began by building a schooner-type dragger at Bay d'Espoir.[6] A year or so later this dragger was bought by a small steel dragger company in the United States.

After a few days in St. John's, I went home for Christmas. Three days after Christmas, Captain Blackwood came to Grand Bank for my extra crew, and he found several other local men before proceeding further into Fortune Bay where the rest of our two crews lived. We left our families and the comforts of home we had always enjoyed each winter to plough through the Atlantic, summer and winter. The next day we all arrived at St. John's, and got down to the real business of deep sea dragging.

The 2nd of January, 1948, our two trawlers, *Challenge* and *Fearless*, sailed out through the Narrows of St. John's harbour, beginning one of the largest fishing adventures undertaken by any Newfoundland company. We steamed all day and part of the night towards our destination on the southwest part of the Grand Banks where we were to look for haddock. Our owners were marketing them to the States.

Everything about this kind of fishing was new to me

There had never been much dragging on the Grand Banks before we started. The English and a few French had been out there a good many years, but we never had anything to do with them.[7] When we were dory fishing I never saw more than ten or twelve French draggers during the whole year. Because we were in over the Bank in shallow water, between thirty-five and forty fathoms, while they fished mainly out in fifty, sixty and seventy fathoms, out on the shelf. They would be out of sight. Many times, on fine days, we could see their smoke, because they all burned coal then, and you could see it for miles out over the shelf. So everything about this kind of fishing was new to me, including the method and where we had to look for fish, especially at this time of year — January.[8] Our dory fishing usually ceased around the first of October, so this was the first time in over twenty years that I went deep sea fishing on the banks in January. There would be plenty of gales, snow and severe frost. But we also knew it would be easier to catch the fish, that there was more comfort on board this ship, and that we had one of the best ships that ever sailed the Atlantic to face it all.

The visibility was excellent at our destination, our two boats were in sight of each other, and it looked like we were alone on the Grand Banks. We set our net for the first time in fifty fathoms, and took it back after towing one-and-one-half hours with only enough fish for the cook to prepare for dinner. Captain Blackwood of the *Challenge* radioed that they had also made a 'blank set'.[9] We hoped, as the saying goes, "that a bad beginning is a sign of a good ending." We fished

Dressing fish aboard the small Burin trawler *Mustang*, c.1945.

for four days and nights, and reported to our owners by ship-to-shore radio at nine o'clock every morning. Neither of us found any fish at all yet.

We went in debt on our food bills

On the fourth morning our owners advised both of us to land the following day. Captain Blackwood landed 125,000 pounds, just enough to pay their food bills, and we landed 70,000 pounds, and didn't manage to break even. It sure was a bad start. Long afterwards, Cyril, our youngest son — only nine years old then — kidded me about that first trip of only 70,000 pounds. We went in debt on our food bill! The food was the only thing the crew had to pay for when dragging. The owners received sixty-three per cent of the catch and paid for everything else. The remaining thirty-seven per cent was divided equally among the crew after the food bill was paid.

During that first trip we all learned something, although we didn't get any fish. Only four men on board had had any experience with dragging methods. We made forty-five sets[10] on this trip and I was very pleased at the progress my men made in such a short time. Captain Blackwood had most recently commanded the *Investigator II*, and had skippered ships for years doing research on the Grand Banks, so he had an advantage over me in this method of fishing. He was one of the smartest men in this work that Newfoundland ever produced. But he pointed out to me that I had an advantage over him from years of banks dory fishing, which he had never done, so we expected a see-saw fight.

We were in port one night and sailed the following morning with a new supply of food on board. After steaming twenty-four hours on trip number two, we began fishing again in deeper water and towed our net for one-and-one-half hours, and had a good set of 15,000 pounds of haddock. After only twenty-four hours we caught as much haddock as we had after four days on our first trip. Everything went well and our hopes were high for a good catch. Our owners were also very pleased with our first radio reports, and, in four days, we bore up for home port with 300,000 pounds of haddock! Our last set was the largest of the trip — almost a whole deck full. After taking the fish from the net, we left for home. It was a pleasant day and we steamed along at about eleven miles per hour, while the men dressed down the fish.

When all the fish were dressed and iced in the hold, the men snugged away the nets by the side of the bulwarks and lashed some with ropes, cleaned the deck, and battened down the hatches. Then a three-man watch was settled. Each watch took his turn cleaning the wheelhouse and shining the brass, and cleaning their own quarters. After this, each man probably had a shower and a good sleep before playing a game of cards. Although we were scheduled to land at nine o'clock the next morning, we were tied up at the discharging dock by 2:30 a.m. and started discharging our trip. At 9:00 a.m., our report was already in the morning paper: "Trawler *Fearless* due to arrive from the Grand Banks with 300,000 pounds of fresh fish at 9:00 a.m."[11]

Now they had all the latest discharging equipment

This reminded me of the time only three years before when I was bringing fresh fish in from the Grand Banks in the schooner *Ronald George* and discharging at Job Brothers. But what a difference! When we landed our first trip then, the company had only small wheelbarrows and the fish had to be hoisted out of the vessel's hold and wheeled away. Now they had all the latest discharging equipment and one of the most modern fish plants in eastern Canada.

By noon the next day our trip was discharged and weighed out at 320,000 pounds. Then the hold of the boat was thoroughly washed and all pen boards taken on deck and washed and scrubbed. By dark, we were ready to sail again with another supply of food and ice on board for trip number three, and I gave the men orders that we would sail by midnight.

The following midnight we began fishing at the same

position we fished on the last trip,[12] and during the next three or four days we had a bumper trip and departed for St. John's with a total of 350,000 pounds. We were back at our home port again within twenty-four hours. Although our first trip was a failure, by January 28th we had landed over one million pounds of fish,[13] and our men shared an average of $300 per man after food expenses, which was considered excellent pay. At this time we received three cents per pound for haddock, two-and-one-half cents for cod, twelve cents for halibut, and two-and-one-half cents for flounder. Haddock was ninety-nine per cent of our catch during this first month.

A tremendous amount of small fish had to be thrown away

The weather on the Grand Banks was good throughout January, except for some frost when approaching within about 100 miles of the coastline on our way home from each trip. The fish were very plentiful during this first month, but a tremendous amount of small fish had to be thrown away in order to secure a successful catch. Sometimes we brought up as much as twenty to 30,000 pounds on one set and were lucky if one-quarter of it was large enough to dress and ice. Every fish smaller than approximately twelve inches had to be thrown overboard. Even a twelve-inch fish will only produce a very small fillet. We kept a record of our catch as it came to the ship and of what was large enough to save, and records were kept for the fishery authorities. We estimated that, for the one million pounds landed this month, over three million had to be thrown back — destroyed — into the sea.[14]

During the first trip in February, 1948, we encountered our first severe winter storm. We were approximately 240 miles from St. John's and had been fishing for one day. There was hardly any frost, but we received reports of zero weather around the land. Her former owners had given excellent reports about her seaworthiness, but this was the first real test of our new ship, and she proved to be the best. We dodged the storms and heavy seas with just power enough to answer the helm.

The only ship in our vicinity was a small Burin dragger, and we talked to each other constantly on the radio. The storm came up on a Saturday, and we dodged the seas until Sunday morning, when I rang the engine room to stop the engine, to see how she would behave, driving broadside to the seas. And she acted very nicely in such a storm. According to radio station reports, the winds were up to eighty and ninety miles per hour around the coast, and we had every bit of that on the Grand Banks.

I knew it wasn't good seamanship to ride out a storm such as this while riding broadside to the seas, but one never knows when the engine will break down, and one seldom encounters a hurricane like this. At 4 p.m., Sunday, I was in the galley having some coffee and our boat was riding well, so I advised the engineer to give us Slow Speed again and head the seas. I had proven we had a good sea boat.

Monday a.m. The storm continued and we received news that the *Isabel Corkum*, commanded by Captain George Follett of Grand Bank, one of the large bank fishing vessels moored up in Grand Bank harbour, broke all her mooring gear and drove out through the harbour and was a total loss. Fortunately, no one was on board at the time. It was the first time I ever heard of a vessel driving out of Grand Bank harbour and becoming a total loss.

Our barometer had been rising constantly in the past few hours and we expected a drop in the storm at any time. By 4 p.m. the wind was decreasing, and at twelve midnight it was below gale force. We stopped the engine and the ship was driving with the seas. By mid-day Tuesday, we were fishing again. Needless to say, I was proud of our boat and hoped I would spend many years commanding her.[15]

It was much like coming out of hell

Early Thursday a.m. We left for St. John's with another big trip — about 340,000 pounds. This method was far, far superior to dory fishing. It was much like coming out of hell and going into heaven. It would take a lot to explain the weight of responsibility and anxiety caused by having twenty or more men at times away from the mothership when dory fishing, and worrying and wondering if they would all return save and sound on board again. If we had been dory fishing in that last hurricane and I hadn't used my good judgement about the barometer, perhaps twenty-two men could have been lost. I was very fortunate and thankful that up to this time I had never lost a man, but I was often anxious.[16]

With dragging, I was on the bridge watching and talking to my men all day long, and when the weather became too stormy for fishing, all I had to do was say, "Snug away everything," which meant don't set the net anymore, and within fifteen or twenty minutes all the men were safe in their quarters and a watch settled. No worries at all.[17]

It was heart-breaking to throw away so many million pounds of small fish

By the end of February we had landed over two million pounds during our first two months dragging,[18] but it was heart-breaking to throw away so many million pounds of small fish. During March we often got more fish in our net than we could handle. The net would come up to the surface with 25- or 30,000 pounds on every set, then the net would burst and the whole haul was lost. At this time of the year the cod came in over the Bank from deep water, and they were also plentiful. But we kept away from it as much as possible. During the past two months we fished in exactly the same positions we fished on our first trip, which had been a failure.

When the cod started to come in over the Bank, about the first of April, we had to shift to shallow water, and that's when we'd run into more trouble. Hard bottom, full of old anchors, rocks and wrecks. Hundreds and thousands of schooners, and especially the big French barks, had fished on these grounds. It would be hard to believe how hard it was, what we fellows used to run into during the first three or four years dragging on these old banks dory fishing grounds. We tore up our nets, lost them, and sometimes lost the whole works — trawl doors and all — where it hooked into anchors and so many wrecks. During the winter months we fished on the shelf of the Bank in water ranging from forty-five to fifty fathoms, where the bottom was mostly muddy, good for towing a net. We didn't even have to put a cowhide[19] on the codend, and our trawl was never damaged except when broken by an overload of fish.

During the middle of May, while I was landing one of our trips in St. Johns, I decided to buy a house and make St. John's our home. We were to come in for a two-week "skeleton refit" during July, and I planned to move my family here at that time. One of our boats would come in for the first two weeks, then the other would follow.

The weeks and months passed and we continued to land big trips. As usual, during the last weeks of June, the caplin arrived and the shore fish in St. John's were so plentiful that Job Brothers' plant could not handle the landings from the two trawlers, so the *Challenge* began her two-week refit from the first of July, and our boat, the *Fearless*, was transferred to Job's plant at Port aux Basques for the next two trips.[20]

It took more than a year to find the hard spots and the good spots for towing

Fearless sailed from St. John's for the banks on July 1st, after landing 380,000 pounds of mixed fish. The fish was well in over the Bank at this time of the year, and our net was damaged quite often during the past few trips because we were fishing in more shallow water and the bottom was very hard in places. It took more than a year to find the hard spots and the good spots for towing. As the saying goes on fishing vessels, "Now we'll find out the men from the boys," and the captain also had to do more calculating.

Before we sailed my owners told me they would prefer me to get a trip of mostly cod to take to the Port aux Basques plant because it was not equipped to handle haddock. This meant, "Be prepared for plenty of rips," as it was always called when the trawl was torn. I said to the mate, "Well, we'll go down now and take a crack at the Bethel Shoal.[21] But it'll be the first time I was ever there in a dragger." Bethel Shoal was at about nineteen fathoms, and we knew from our dory fishing years that there was always a lot of cod there, especially at one spot about ten to fifteen miles around, which was a good place for the large codfish. I had planned to go there when the time came because we got a great many trips of cod there when I was landing it fresh in St. John's from the *Ronald George*. We steamed twenty-eight hours from St. John's and began fishing on Bethel Shoal.

You would see those big fish floating away

The cod were there alright, but we had our net out only fifteen minutes when it hooked the bottom and had to be taken back. We were greeted with wreckage from an old ship and an anchor weighing seven or 800 pounds, and large codfish were floating on the water where our net was torn. But we saved 10,000 pounds from this first haul. Every time we towed our net I had to have my hand on the telegraph and at the least bit of pluck at all you knew the net was torn. But if we could tow the net ten minutes, we would haul them up and it would be so full of those big, large codfish as it could be. And perhaps the next time we brought her up you would see those big fish floating away where the bottom was torn out of the codend. We kept going like that with plenty of fish and plenty of torn gear.

One evening I was talking over the radio with Roy Russell, one of our owners in St. John's, and we had only about 100,000 pounds aboard. I turned in at about twelve o'clock midnight for a rest, because I had been up all day and all night, and at eight o'clock the next morning our big deck was filled from stem to stern, so full as you could ever get it, with big codfish. The men were at it all night, putting down whatever they could.

After fishing there for four days and nights, we departed for Port aux Basques with 270,000 pounds of mostly cod, and a few thousand pounds of halibut. They were paying two-and-one-half cents per pound for cod then. On our next trip we landed another 250,000 pounds of cod and landed it at Port aux Basques.

We got all the grounds clean

When we fished the hard bottoms we used to bring in as much as eight or ten old, rusty anchors on the head of our boat from each trip, and when we got her in the deep water of the Narrows of St. John's harbour, we dumped them all overboard. That's what we were up against all over the banks, the first years we went dragging. The two or three dragger skippers up at Burin — Captain Bobbie Moulton and Captain Arch Broydell — and Captain Baxter Blackwood and I, we fellows cleared the ground for the skippers who went dragging eight or ten years later. Bethel Shoal, in particular, we had as clean as your living room. But we got all the grounds clean, and there's hardly a place out there now that you can't go dragging.

Our owners heard that I was moving my family to St. John's

Before we sailed on July 1st, our owners heard that I was moving my family to St. John's and gave me permission to touch in at Grand Bank on my way from Port aux Basques to pick up our belongings. On July 10, 1948, after our last trip was discharged, we left Port aux Basques on our way to St. John's via Grand Bank for our refit. I didn't tell my family that I was calling for them until we were coming around Grand Bank Cape the next morning, when I switched on the transmitter and said, "Get ready, Ruth. We'll soon be there. We're coming around the Cape now." I knew they would be listening for me to talk to the other boats on their radio receivers.

As quick as we tied on, at eleven o'clock that morning, about fifteen or sixteen of the nineteen men aboard went with me right up to the house. And we never packed a thing, just picked it up and brought it right down to the boat, from the piano to the smallest thing we had. By four o'clock we had everything aboard, and we sailed for St. John's at eight that evening.

Coming down, there wasn't a motion of sea all during the night and the next day. What a pleasant trip! On Sunday morning, Florence, our youngest daughter, steered the ship along for thirty miles and was very proud of her accomplishment. Our cook also had all of his family on board, taking them to St. John's for a holiday, and all the passengers were thrilled over their pleasant passage. We arrived in St. John's just before dark on Sunday, July 12, and all of our men, except a skeleton crew, went to their homes from Grand Bank to Fortune Bay for a two-week holiday, while the boat was having a refit. The *Challenge* sailed for the banks the day before we arrived as she had finished her refit.

My wife and I were both very grateful and thought it was wonderful that Job Brothers gave us this opportunity, and it was even more appreciated when I decided to take all the family with me on the boat. And Job Brothers never charged me five cents. At the time it would have cost $600 or $700 to move everything down.

Ruth felt right at home

I had some renovations being done on our new house, but it wasn't quite ready when we arrived, so the whole family stayed aboard the boat for two nights. Although I knew the children had been down to St. John's for two months the summer before, visiting family, I couldn't get over how Florence and Roland went ashore and to the theatre as if they had always lived in the city. Ruth felt at home because her mother came from St. John's and it was part of her home.

(And Ruth remembered: I didn't want to come. I don't like moving. My mother went out to my home, Pool's Cove, to teach, and met my father, and we were all born and raised there. I visited relatives in St. John's at times, but we were strangers in a way. But the children didn't mind it too much. Right away, they wanted to go to the Regatta, and I said, "You might get lost down there in the crowd. What would you do if you got lost?" And little Cyril said, "I'll tell 'em to take me to 205 LeMarchant Road. Me mother will pay for me.")

Grand Bank always meant a lot to me

Every summer for the first three years while in for our annual refit, I went to Grand Bank for my vacation. My wife and children went back there after school closed, and lived in our house. Then a fellow there wanted it badly, so we decided to sell it. I always regretted selling it, because it was a lovely new house, well built, and only five years old, and we didn't have to then. I wouldn't care if I only had it to look at when we would go up there. Grand Bank always meant a lot to me. It was there I went for my first year, going to the banks, and where we went to live after getting married. And all our children were born and got most of their education there. We had, and still have, a lot of friends there, and it wasn't easy to leave.

Refitting work began on our boat and we were busy for the next weeks. Meanwhile, the *Challenge* landed two good trips, about the same amount as we landed at Port aux Basques while they were in refit. Our two boats had had excellent results since we started on the 2nd of January, landing almost four million pounds each,[22] but neither Captain Blackwood nor I expected to land that much during the rest of the season. About this time, midway through our 1948 voyage, our owners renamed our two boats: The *Challenge* was now the *Blue Spray*, and the *Fearless* was renamed the *Blue Foam*.

On July 28, 1948, we sailed from St. John's for the banks. The haddock had practically disappeared since the last part of June and the fishery scientists in St. John's couldn't tell us why or to where. So it seemed we had a lot of research to do on our own to find good bottoms to drag our nets on. Captain Blackwood had an advantage over me in this, because of his research experience in the *Investigator II*. Nevertheless, we both expected good trips of fish.

Our owners ... preferred that we find perch and flounder

Our owners continued to tell us the cod markets were poor, and they preferred that we find perch and flounder. On our first trip after refit our men lost a lot of sleep as we fished hard bottoms and tore our net on almost every set. In four days, we had caught enough to land, and arrived at St. John's with 280,000 pounds, mostly cod. During August we continued to catch bumper trips of cod, which our owners were splitting and salting. We fishermen were just as glad to catch cod as haddock, because it was only half a cent less in price and much larger for dressing. But we wanted to supply our company with the species they needed for their markets.

September was more satisfactory for our owners. We located some flounder and landed bumper trips of cod and flounder. On the last trip that month we landed a full load — our hold carried 385,000 pounds of iced fish, and 10,000 pounds on deck. We did quite a bit of research during October, looking for other species, but almost always ended up on the codfish grounds. Despite much gear torn by old ship wreckage and rusty vessel anchors, we managed to keep our landings pretty high. We hoped to locate some haddock by this time, but it seemed that we would have to wait until the winter season again.

Our boat was to be docked in November at the C.N.R. drydock to have its hull inspected, cleaned and painted, but a strike there forced us to go to Sydney, Nova Scotia. It took the best part of a week to get docked and off again, and on our next trip that month we couldn't find any suitable bottom to tow our net on and we had a tremendous amount of torn gear all the time. It was our worst trip of the year, except for the first.[23] Our last two trips in November were mostly cod and flounder, and our catch this month was down.

Our sister ship's main engine broke down on one of her November trips, and when her captain called to report her troubles we were about 100 miles from them. They were about 280 miles from St. John's. Our owners immediately ordered us to proceed and take them in tow for home port. We were alongside them within ten hours and took the boat in tow. About four hours later we arrived in St. John's. Two weeks later the *Blue Spray* was on the fishing grounds again with everything working well.

December. On our first two trips this month we landed two loads of cod and flounder, and on the third we landed a full load of haddock. After discharging, cleaning the hold, and taking a load of ice aboard, our men went to their homes in Fortune Bay for Christmas, which was about four days away. If all went well, we hoped to land one more good trip this year, so at midnight on Boxing Day, we sailed again and steered our course for the southwest part of the Grand Banks. There we found haddock plentiful and loaded our boat in three days, and arrived back in St. John's in time to celebrate New Year's Day.

We landed the largest amount of fish every caught

The owners, crew, and I were all pleased with our first year's work. We landed the largest amount of fish ever caught in one voyage in Newfoundland history — ten million, 500,000 pounds of mixed fish.[24] Our men shared a total of $3,600 each for their year's work, after their food expenses were taken out. There was no mistake about it. Deep sea fishing had changed in a way none of us expected in only a few short years. It was now a full-time operation, winter and summer. But it was really worth it. And there was no comparison with what we did in the past. The comfort, way of fishing, the food and many other luxuries made it just too much to believe.

I had to call men out . . . to beat ice off the ship to keep it from sinking

It wasn't always easy to secure a trip of fish during the voyage because of the wrecks and old anchors that damaged our nets, and lost nets and 1200 pound trawl doors. And many times, coming to land with a load of fish, we ran into fierce gales and zero-degree frost, and I had to call the men out of their berths to beat ice off the ship to keep it from sinking. Yet we had been fortunate in not having one engine breakdown, and most important of all, no injuries — which I dreaded most of all, especially because our men were unfamiliar with working the winches, handling the 7/8 inch cable used for towing warps, and the heavy trawlers and other gear.

The iced up deck of a Job Bros. side trawler at dockside in St. John's, c.1950s. A crew member poses with wooden mallet used to clear deck and gear of ice when at sea.

If the destruction of our small haddock kept up, there would soon be none left

I had a wonderful crew of men that made this successful voyage possible, and I've always given them the highest praise. Except for my three engineers and the four men who had been dragging before, the others were all with me on the *Ronald George*. They were all eager, pleasant, hardworking men, and I was very pleased to see how quickly they learned everything. Before our voyage was over nearly every man could make up the trawls, mend twine, and splice wire as well as the men with previous experience, so I expected our deep sea fishermen had a good future for the next twenty years at least. One thing I feared, however, was that if the destruction of our small haddock kept up, there would soon be none left.

We both liked to be the top man when we came in

Baxter Blackwood and I were neck and neck for ten years before he left dragging. We didn't tell all truths to one another. Oh, no. You couldn't. But when he was here in St. John's, there was no brother or anyone else closer than he and I were. But I know he used to tell me lies and I used to tell him lies, trying to beat one another. We wouldn't do anything to hurt anybody. If he was out there going around a couple of days and didn't find any fish, and I was on fish, I would tell him and he would do the same for me.

We both liked to be the top man when we came in. I remember one night in January. It was blowing and snowing, and we were talking to each other over a ship telephone. We were fishing away out on the Grand Banks. Around 10 or 11 o'clock Baxter says, "Tis getting pretty bad now, Arch." You couldn't see anything. A thick snow storm. And I says, "Yes, boy, not going to fish much longer, going to take it in on deck," going to quit fishing. Well, you have to give up when the wind gets us (to around thirty-five or forty miles per hour or more, depending on sea conditions). You can't fish. And

Millions of Fish

he says, "Yes," he was going to do the same. By and by he called: "What time are you goin' to give up, Arch, boy?"

"This is the last set, boy. When we get it in I'm going to tell the men we're going." "Yes," he says, "I am too. Matter of fact, we got ours on deck now. Just taking in the last of the fish."

So when we got the last of the fish in the squall wasn't so bad. I said, "Put them over again" and we fished the whole night. We never gave up for the night. Daylight came and we started over again. "Pretty bad, Arch," he said, "pretty bad". "Yes, boy, not fit for a dog." But I never told him we were fishing.

On the third day of the New Year, 1949, we sailed from St. John's again for our first trip of the season. As we were going out through the Narrows, the *Blue Spray* was coming in with her first trip of the year and she looked well fished. We saluted each other as we passed.

Fish continued to be plentiful and many bumper trips were landed during the first four months that year, and by the last of April we had landed another five million pounds, ninety-five percent of it haddock. As in last year's catches, however, we dumped back into the sea — destroyed — about three times that much small fish — fifteen million pounds.

Through May we got large trips of mixed fish — haddock, cod and flounder. Then, on the first of June, our owners had us take salt on board our boats and prepare for salt fishing, as the markets for fresh codfish were zero. We also took some ice and planned to fish for flounder during the last twenty-four hours of each trip. Our men had been salt fishing most of their lives, and were excellent splitters and salters, so the change didn't bother us.

During the first four days on our next trip we took aboard all the cod we could handle until we had space left for 200,000 pounds of flounder. Then we steamed from the cod ground to the flounder ground, and the flounder were there waiting for us. We secured our 200,000 pounds in twenty-four hours, iced it down, and left for St. John's with another full load. We discharged our cargo, took on another supply of ice and salt, and returned to the same positions and loaded our boat again. We continued to bring in full loads like this until the 12th of August, when the owners had to stop salting the cod down on the boats as they were to be split and salted when landed. The flounder market improved and our trips consisted mostly of it until November.

The markets were fair for ocean perch or redfish

Our owners kept reminding us that the markets were fair for ocean perch or redfish, but until then no one had discovered any in the Grand Banks. In fact, none of us wanted to lose any time searching for this species because we were getting all the cod, flounder and haddock we could handle, each in its season — haddock in the winter and spring, and cod and flounder the rest of the year. During the past few years Fishery Products' Burin boats were taking loads of redfish in Hermitage Bay, and we spent a few days there once in the *Blue Foam*, but it was too close to land for us to have any interest, so we never returned.[25]

Groundfish — haddock, cod and flounder — were very scarce during the first part of November on the southern part of the Grand Banks. There were just the two of us there, about thirty or forty miles from the edge, on this occasion, one boat (*Zerda*), commanded by Skipper Bob Moulton, and ourselves. Captain Moulton and I decided to go out from shallow water, about forty-five fathoms, to a depth of 180 fathoms on the shelf of the Bank to search for redfish. We had no fish finders at the time.

What happened when we hauled ... back was unforgettable

After driving to our intended depth, we both shot away and set our nets at the same time, only a short distance apart, and planned to tow one hour if we didn't hit hard bottom and damage our net. What happened when we hauled them back was unforgettable. The two trawls were packed so full of fish that they broke open as quickly as they surfaced and the whole catch was lost. That began the discovery of redfish on the Grand Banks. No time was lost repairing our gear and setting again. By our third day of fishing at this position, we had a full load and left for St. John's. Our owners were pleased to learn that we had discovered another species of fish on the Grand Banks to add to the market. During the rest of November we brought in full loads of ocean perch. Although it was out on the edge, and you wouldn't drag up any anchors, the redfish bottom was up and down, and we were to spend a lot of time feeling out the level bottom from the humps and mountains.

The haddock and ocean perch on the Grand Banks are both on the small size compared to those found on the other fishing banks. The redfish are smaller than those in the Gulf of St. Lawrence and Hermitage Bay and other places where they concentrate. They are large enough to produce a satisfactory fillet, but we much preferred to see some larger ones come along. In December we brought in four loads of ocean perch and flounder. The position where we secured our redfish was only three-quarters of an hour's steam from where we found all the flounder we could handle.

As we fished for redfish, we found that as soon as the sun set, they left the bottom and could not be reached with the bottom trawl. This was because what they fed on left the bottom and the fish followed. Many times, on the last set before dark, our net was as full as it could be, so we'd set again for curiosity, but with no success until the next morning at daylight. Many times we wanted only one more set of, say, 20- or 30,000 pounds to fill the hold to the hatches, but had to wait until morning. So we began to fish the ocean perch during the day, and steam in and fish the flounder ground by night.

There were always other men waiting for a berth

A few days before Christmas we arrived at St. John's with a full load, discharged our catch, and all of our men went to their homes for Christmas as they did last year. A few asked my permission to stay home with their families for a month, but I had to promise that they could come back again. They could be replaced easily, as there were always other men waiting for a berth.

December 27th. We sailed again for the redfish ground, approximately 300 miles from St. John's, and fished three days, averaging 115,000 pounds each day, and arrived back at St. John's on the second day of January, 1950, with 345,000 pounds on board to end another very successful year. It is hard to believe, but our catch in pounds was about the same for each of these first two years at dragging — 21,000,000 pounds altogether.[26]

We were all pleased with another good year's work. Everything went along pretty smooth, except for a few storms and frost, and some damaged gear, and we were also fortunate in not having any accidents.

On January 4, 1950, we sailed on our first trip for this season. The boats already on the banks had not found any haddock so far this month, and they were all on the redfish ground. We steamed thirty hours from St. John's and began fishing for redfish, and, by the first of February, we landed 1,200,000 pounds. By then the haddock were moving in from deep water.

From February through May we brought in bumper trips and broke our own records many times. And during June and July we had big trips of cod and flounder.[27] In August we located haddock in abundance feeding on caplin on the Bethel Shoal, at nineteen to twenty fathoms, the shallowest water on the Grand Banks, except for the Virgin Rocks. As I said before, we caught many trips of codfish on the Bethel Shoal when we were dory fishing, and Americans and Nova Scotia vessels would get a few large trips of halibut there at certain times of the year. We succeeded in getting two large trips of mostly haddock, and some cod, for a total of 600,000 pounds from two trips. Once again, there was plenty of damaged gear from old anchors and wreckage.

The large whales, and sea birds, were numerous, in the thousands, in the small area we fished in on these two trips. They fed on the caplin bait. Most of the time we made very short tows, as short as ten and fifteen minutes when the fish were so plentiful, and then we'd stop our main engine and dress the fish. As we lay there, the whales came alongside by the hundreds and slapped their long fins by the side of the boat. It was really a pretty sight. But when the caplin moved away, so did the whales and haddock.

I'll never forget how hurt her captain sounded

We brought in large trips of cod and flounder during September and October, and flounder and mostly redfish in November and December. On one trip we were fishing about 330 miles off from St. John's, down on the southeast edge of the Grand Bank, two or three miles from where the continental shelf dropped off into over 1,000 fathoms of water. It was a Sunday and, well fished with 315,000 pounds stowed away, we had just battened down our hatches and were steaming towards home for about an hour. I had my radio on 2182, the emergency channel, when I heard this ship calling, "Coast Guard. Any Coast Guard." Then he began calling "Argentia. Calling Argentia." His set seemed powerful to us, but he was apparently getting no answer. I couldn't understand why somebody wasn't picking it up, so I went back to him and said, "Captain, I hear you calling for a Coast Guard cutter, and Argentia, and I wondered if there is anything I can do for you?"

It was the Gloucester trawler *Gudrun*, and I'll never forget how hurt her captain sounded when he said, "My God, Captain, while I'm talking to you I have a man aboard, I don't know whether he's dead or alive. I just arrived from Gloucester and it's my first trip on the Grand Banks, and we was just putting our trawl doors out for our first set when the man got caught in the main winch, and I believe he's almost mangled to death. That's why I'm trying to reach somebody. I'm a stranger to this part of the world and wondering how near a Coast Guard station is to us? I know Argentia is an American base, but I can't seem to get any answer from there." It was about 300 miles to Argentia, and about the same distance to St. John's.

I told him we were just leaving for St. John's and if we could help, and it's O.K. with him, I would change course and go to him with all the power we had and take his man and rush him to St. John's as fast as we could. The captain didn't know how to thank me. Before we finished talking, I got a bearing on him with my little DF — direction finder — and had my course changed, and we went so fair for him as if you had drawn a line. In less than two hours, shortly after dark, we were alongside his boat. Fortunately the sea was as calm and smooth as one could see, especially in November.

While I steamed up the fifteen miles to the *Gudrun* I got in touch with my owners and told them what I was up against, and they gave me the O.K. to do whatever was necessary. That was the first thing you were supposed to do — get in touch with the owners, to do the right thing. Then, if anything should happen, you had no problem; not the way it was when I was in the *James and Stanley* that time I had to jettison that oil.

Earlier that same year, just after we had our refit in St. John's in late July, we were steaming about sixty miles from St. John's on our first trip to the fishing grounds when we sighted an empty dory drifting on the water. We changed our course and picked it up. It was a small, one-man Portuguese dory. I said to the men, "Lash it on top of the after deck. You never can tell, she might be useful for something." And that was the dory we used that night instead of our big lifeboat that took about twenty men.

He was a real brick

Our men had the small dory ready to put in the water by the time we reached the *Gudrun*, and the mate and two men got in and went over. Within minutes we were taking the injured man on board our boat and it was really a hard sight to see.

The man, about forty-five years old, had been operating the main winch controlling the heavy towing gear when both of his hands became entangled in the towing warp. The huge drum whirled him around three times before the other fishermen could help him. When they freed him, one arm was hanging on by torn shreds of flesh at the elbow, and the fingers on the other were severely crushed.

After my owners confirmed what I was doing, they immediately contacted a doctor, and we got together on the ship-to-shore radio. I had some morphine tablets in our medicine chest, and he instructed me how often to use them. We kept in contact with each other every two hours that night. The injured man was in severe pain when we took him on board, but his pain seemed to ease after taking the morphine, and he never moaned that night. Our men, two at a time, stood by him all that night, constantly giving him cigarettes that also helped deaden the pain, and he talked to them all the time. He was a real brick. If you didn't know how badly he was mangled up, just his two arms, you wouldn't know he was hurt.

By daylight the next day I could see he was losing a tremendous amount of blood, and I feared he might die before we reached St. John's, so I contacted the authorities at Fort Pepperell, the American Air Base, and asked them to have a seaplane come out to meet us. They assured me they would do everything possible and advise me within a few hours, and asked me to keep our radio open all the time. Within a few hours they advised that the sea was too choppy to send a plane, but in the meantime, if I thought of anything else that might help, not to hesitate to ask them. The weather remained excellent and the *Blue Foam* did her very best.

By noon St. John's was still about 130 miles and twelve hours away, so I asked that a boat with a doctor on board be sent out to meet us. If a doctor could meet us even twenty miles from land, it might save the injured man's life. They agreed and sent a boat with two American medical men aboard, and we met them about fifty miles southeast of St.

John's at eight o'clock that night. The sea was very calm and they had no difficulty coming alongside to climb aboard. The medical men cared for the injured man until we reached St. John's at about 1 a.m. An ambulance immediately rushed him to hospital where his fingers and arm were amputated, and he survived.[28]

Captain Johannssen, the *Gudrun*'s skipper, resumed fishing on the edge of the Bank just as soon as we took his injured man, and he was ready to leave for Gloucester by the time we discharged and returned to the Bank.[29]

On December 24th we arrived at St. John's on our last trip before Christmas and, as before, after our fish was discharged, the crew went home for Christmas. We sailed again for the banks on December 28th. The weather was stormy on this trip and we were longer getting our load. On January 5, 1951, we arrived at St. John's with our last 1950 trip, another load of 350,000 pounds of redfish.

Our 1950 voyage was large — 9,000,000 pounds — but not quite as much as on the previous two years.[30]

What we destroyed was a drop in the bucket

So many haddock were being destroyed, and the number of boats fishing on the Grand Banks was increasing. There were already several trawlers from the United States visiting us, catching redfish, and we saw quite a few Spanish twin trawlers, two boats towing one huge net, this past year.[31]

We saw these large Spanish trawlers take back their nets, and have haddock enough on one side of their boats that they listed so badly the water came in over the rails, even on a very smooth day. Out of from 25- to 40,000 pounds of fish on each haul, they would get only 5,000 pounds big enough to split and salt, and put in their hold. All the rest were gone, dumped overboard, and within an hour, they were towing along beside us again without one fish and hardly a man on deck. This went on year after year. There were two Spanish trawlers on the Grand Banks the first year I started dragging, the *Santa Elisa* and the *Santa Maria*. Before the second year was over there were four or five, and about a dozen within three or four years.

Our few Newfoundland trawler skippers complained among ourselves all the time and to our owners when we came in about what we were destroying. Our owners could see from the records we kept. And we lost a lot at the plant. Our "weighback" — fish the plant thought undersize (plus trashfish and ice) — in 1951 alone was about 1,000,000 pounds! But what we destroyed was a drop in the bucket compared to what the foreign trawlers did.[32]

The new technology meant fewer jobs

Dragger fishing began a new way of life for me and the Newfoundland fishing industry, but I often thought about our dwindling deep sea dory fishing fleet, which by 1951 was down to the last ship.[33] One could see that the new technology meant fewer jobs and more unemployment, and our Newfoundland deep sea fishermen knew no other way to make a living. I wondered what would happen to them. I was a successful man with more comforts at home and at sea than I ever had before, yet I couldn't help but feel some concern and sadness for my former fishing mates. And draggers would certainly deplete the waters for both the deep sea dory fishermen and the shore fishermen. There would still be a lot of men depending on deep sea dory fishing and shore fishing in Newfoundland. All I could do was hope to help as many as I could to become as lucky as I. After thirty years of depending on the same type of livelihood, it was not easy to forget.

Speaking of new technology, by January 1951 our boats each had a direction finder with an effective radius of about fifty miles, an echo sounder to give us the depth of water, and, since early 1950, a Loran, which gives an accurate position from the southern parts of the Grand Banks — over 300 miles — to the coast of Newfoundland. Radar and a fish finder — to locate the fish and give us its depth — were promised us in the near future, and we were anxious to see

274 Voyage to the Grand Banks

Portugese brig on the banks, c.1946. Note pattern of repairs to sails, a sailing ship's individual signature when seen at a distance.

these new instruments. Our radio set and receiver was very powerful, and at times I could even talk with vessels as distant as Turks Island and Barbados. I remember during the first year I was dragging that Ruth often had the radio on at home because she could hear us talking to one another out on the banks. I knew most of the sets on the schooners were closed down by about twelve o'clock at night, so I called Ruth every night just to say, "Good night. I guess you got them snugged away by this time." Every night she tuned in and would know where we were and that everything was alright.

He said I wasn't very loyal to my company

Near the end of our first trip in January, 1951, the trawler *St. Richards*, from Burin, sent out a distress call. She was on her first trip on the Grand Banks and had 150,000 pounds of rosefish (redfish) in her hold, and was just completing a tow when she suddenly rolled over and a big sea entered her open hatches. The crew tried to get her on an even keel by cutting her net loose, but the fish aboard had already shifted. Then the net became entangled in her propeller, so her skipper, Arch Broydell, was unable to bring her up to windward. He called for help and we rushed up to him, along with the draggers *Zibet* and *Blue Spray*. When the *Zibet* reached him, Captain Broydell transferred his crew to her and, within minutes, the *St. Richards* capsized and sank. Her cook had a heart attack and died just before the rescue, and was the only casualty.[34]

Back in St. John's the next day or so, one of my owners complained that I hadn't put her under tow and tried to salvage her. He said I wasn't very loyal to my company. And I said, "I might have salvaged her, but she would have capsized later for sure, maybe with a full crew. I would have been loyal to my company. But I wouldn't have been a friend among all the fishermen of Newfoundland." She just wasn't seaworthy.

They're gone

I picked up Captain Johannssen of the Gudrun on our radio just outside the Narrows as we were coming down for our first trip in January. He was on his fourth trip ever to the Bank.[35] We chatted off and on until we reached the same position down on the edge, about 330 miles off, where we knew the fish were abundant, and we started fishing together on Sunday morning, January 7, at exactly the same moment. Almost close enough to touch one another, we fished there side by side, tow after tow, until early the next Sunday, when we both departed at the same time. The *Gudrun* was well fished, with about 260,000 pounds, mostly flounder.

The day before, while waiting to take back our nets between tows, my men went up on the head of the boat and took off their shirts with the sun blazing down on them. The water was so warm it made you feel like jumping overboard for a swim. There was never a better day in July. Yet we knew from the low glass and radio report that there was a storm then around St. Pierre Bank, and that we were going to run into it some time the next day. There was no way to get around it. (At the same time on shore, a fierce blizzard hammered St. John's and the eastern seaboard.)

By four o'clock the next morning we were coming under full steam and she was doing everything, almost turning inside-out, when the man on watch came into my cabin and said, "Skipper, I guess probably if you was out, we'd slow down now." I went to the bridge then and put her right down to Slow Speed, just for enough headway to answer the helm, and she dodged away from her course. We punched along and dodged in the face of an eighty mile per hour gale and mountainous seas all that day and night until some time the next day, when the wind moderated enough to let us get up speed again.

At eight a.m., Sunday, another skipper, on a Lunenburg dory trawler, informed me that the *Gudrun* had sent out a short distress call four hours before, that she was sinking. By that time, she would have travelled about seventy miles from where we separated. Numerous ships and boats, and planes from Argentia and Nova Scotia, searched for her, but never found a sign of her crew. Two or three Lunenburg boats were only three or four hours away up to windward, and they swung right off before the wind and sea, ran to their last position and searched all day, but there wasn't a sign of a thing.[36] By four o'clock that evening, we knew there was no sign of life, and I said to the boys aboard, "They're gone. Their lifeboats couldn't stand it. There's too much sea." But one lifeboat was picked up, empty, about a month later.[37]

For several trips after that, when we landed in St. John's, someone from Gloucester would be waiting to meet and question me about the *Gudrun*, because we were the last to see her. And I had a very sad letter from Captain Johannssen's wife, asking me to repeat his last words. She told me that the *Gudrun* was named after his daughter, and she was heartbroken. Almost every time Captain Johannssen came on the radio he was thanking me for bringing his man in and saving his life. When he left for Gloucester that last Sunday morning, he said, "So long, Captain. And thanks again for everything. I'll never forget so long as I live what you did for us that night." And I told him, "Look, Captain, I did exactly what you would have done if you were there."

January, 1951, was also a very surprising month because the weather on the Grand Banks was about as warm as it could be in the summer. The temperature even on the western and northwest parts of the Bank was so mild when we came to land that there was no ice at all freezing on the boat. But mild weather has its advantages and disadvantages. Haddock was so scarce that you would think you were not fishing the same ground as in other winters. There was absolutely none at all, so all of the Newfoundland trawlers concentrated on redfish until the month of May, and many bumper loads were caught. Our boat alone landed approximately 4,000,000 pounds in these four months.

American trawlers ... claimed that the redfish were depleted on their grounds

Beginning in May we began catching cod and flounder. We had a bit of company that winter and spring; quite a few American trawlers were gathering on the southern edge of the Grand Banks, fishing on our grounds for redfish, and sometimes flounder. They claimed that the redfish were depleted on their grounds, and in my opinion, that would not be good for our Grand Banks. I had already talked with some of the top fishery scientists in St. John's on several occasions. Once, in particular, the man in charge of the Fisheries Research Laboratory there and I were having coffee in the galley aboard our boat, and I told him, "We fishermen are very much concerned about other boats coming from other countries, destroying our small haddock, and now catching up our redfish." I'll never forget the answer he gave me. He said, "Skipper, for the next hundred years, someone will be seeing what you are seeing now on the southern edge of the Grand Banks, millions of redfish."

As time went on, our catches cut back every year

I said to him, "Sir, I cannot and will not agree with you. Within the next ten or twelve years, there won't be enough haddock or redfish on the grounds to fish." There was no haddock on the Grand Banks in 1951, perhaps because of the mild winter. That was our fourth year. But as time went on, our catches cut back every year, and by the thirteenth year, my last year on the *Blue Foam*, our landings were down to 6,000,000 pounds.

Around the middle of May, Catherine, our oldest daughter, graduated with her degree in Music from Mount Allison University. I wanted very much to accompany my wife to her graduation, but could not get leave to do so. We were proud of her success during the last three years, and she looked forward to teaching violin and piano at Prince of Wales

College in St. John's at the opening of the school year in the fall.

During May and June we continued to bring in bumper trips of cod and flounder, but all of us found it hard to forget about the scarcity of haddock that past winter. Our owners were very concerned because their markets were so dependent on it, and we all kept our fingers crossed that the coming winter would be better.

We were in St. John's most of July for our annual refit. Our crew and I looked forward to that very much, as it was the only summer holiday we got during the twelve-month period. Fishing resumed again after our three-week refit was completed.

During August, September and October we found cod and flounder plentiful and landed many bumper loads. In November we began concentrating on redfish again.

Chapter Twelve

Epilogue: The Voyage Ended

Skipper Arch suddenly passed away on the morning of 27 June 1976. I had spoken with him on the phone only the night before. He told me he would be leaving the next morning to fly up to Halifax, to visit his eldest son, Roland, near Dartmouth, for a few weeks holiday. He was taking his scribbler along to continue writing about his years on the *Blue Foam*. We would get together again later in the summer.

I was bearing up for the door shortly before noon that morning when the phone rang. It was Arch's younger brother, Cyril, with whom I had not spoken before. He told me that Arch had had a heart attack early that morning and later died at the Grace hospital near his St. John's home.

The news nearly shocked me speechless, but I managed to express my sympathy. I would stop by to see the family later. My youngsters were waiting outside for a promised canoe ride. I went on with them and kept my thoughts to myself. The "old man" was gone. I would have sailed anywhere with him, and I felt almost like I had lost my own father.

Later that night I came up the stairs of his home. Catherine, his eldest daughter, was standing at the head of the stairs, very upset. And she said, "Oh, Andy, the book wasn't finished. And it meant so much to him."

I recalled our last meeting there. He was still very depressed about the recent loss of his old friend and brother-in-law, Captain Clar Williams, and he reminisced about his father's death at home in Anderson's Cove many years before. His father had felt poorly after eating dinner and barely managed to reach the couch. Then he said, "I don't think I'll make it over the barrens for the caribou season this year." Soon he lapsed into unconsciousness and his breathing became laboured. He remained like this for several days before he died.

And Arch told me of how he and Clar often talked of death and how neither was afraid of dying. But a long, lingering death was not to their liking. Let it be over quickly. Both had their wish.

I reflected over what course to follow with our book undertaking. We had not covered as much ground as I had wanted. Yet it seemed that what I had in tape recorded interviews and written form justified continuing the project to its conclusion. Now, writing these final lines, I remember how Skipper Arch spoke of his own work, year by year. At the end of each voyage, he always said, "I ended the year." And never that he "left the vessel."

In late March 1976 I asked him if he had ever thought of giving up the sea and finding work on land before two heart attacks forced him to stop dragging in 1960, his thirteenth year as master of the *Blue Foam*. His answer:

> No. I can't say once in my life. Fishing was about the only thing a man without any education could do at the time. I have been discouraged out at sea many times. That's only human. I could easily have given it up that first year, when I got so seasick. But, if I'd given in to that seasickness just once, I would have been finished. If I had just got in my berth and said, "I can't go today," I would never ever have gone to sea again afterwards. I still think of what would have happened if I had got sick bad enough, if I hadn't fought it. And I really did fight it. And if the captain had had to carry me to land... And it was even worse than what I said about it.

Epilogue: The Voyage Ended

Arch was a finisher, not a quitter. And optimism dominated his approach to life and people.

It is difficult to generalize about his generation's values without risk of contributing to an oversimplified, idealistic and distorted portrayal of a complex people. But his story of hard struggle and determination does illustrate qualities of character noted by others familiar with outport Newfoundlanders of his era. For example, historian Eric Sager (1981:110), points to Norman Duncan's stories of outport life around the turn of the century, who finds an . . .

> . . . enormous capacity to endure hardship, the refusal to despair when faced with an unyielding ocean, the failure to become brutalized even when hunting and killing to live, the rejection of 'dumb fatalism' when surrounded by overwhelming natural forces, and the dogged persistence which ensures survival.

Arch's saga reflects similar qualities and values among his generation. They are somehow some of the results and imperatives of lifestyles followed in fishing outports, and the gruelling deep sea fishing labour regime they followed. And what of the skipper? How did his command responsibilities shape what he was to those who knew him?

Banks dory fishing under sail schooled Arch in an occupation of frequent danger in an era of harsh economic and working conditions. Many banks skippers are remembered as extremely arbitrary and unreasoning. As his brother Stan' observed,

> "A man wasn't valued no more than a dog them times." (And those in command), . . . how stuck up they used to be. They were always right, no matter what. Just say one little thing and you lost your chance next year on that vessel and others. Had to bite your tongue all the time."

One might expect such skipper behaviour and treatment of crewmen to engender a similarly harsh, dictatorial approach in the next generation of masters. Indeed, by both

Arch's and Stan's recollections, Arch became more absolute in command rather than risk losing control of his crew. Skippers were individuals, but their fish killing was their responsibility.

To Myril Herridge, who sailed with him on the *Pan American*, "Little Dan," as Arch was often referred to by the men, seemed to take things so easy. Yet he was a "gutsy go-getter." To other men, he was simply "Tough. When he gave an order, you jumped, or you took your clothes ashore next time in port." Men did lose their chance to sail under him for various reasons. But all banks schooner and, later, dragger skippers experienced some crew changes from trip to trip, from voyage to voyage.

Arch told me that, before unions came along, skippers probably had too much power, were too unchallengeable. Perhaps he became a reasonable master. He recalled to me an occasion when dory fishing, a day when he decided to put his men out and they refused to go!

> One time it happened to me and I give the men credit for it. I said, "That's good, boys. You fellows have to go out there." They'd go down below then and turn in, get in their bunk, turn their backsides out and go to sleep for the day. What could a skipper do? He couldn't put them overboard, one man against twenty-odd. But I wasn't that hard. As quick as we got snugged away, I went down, talking with the men just as if nothing had happened.

He believed the event was uncommon in those days. Other skippers may have reacted similarly — on occasion.

Bill Pope, once Arch's mate on the *Blue Foam*, described him as, "eager, all go, no slack. But not that hard on his crew. He'd give you a break when he could." He was a stickler for details, careful preparation and planning, and quick to measure a man. Cyril, his youngest son, observed: "He could look at you, talk to you ten or 15 minutes, and he knew if he liked you or not." He had no patience for "sea lawyers" or boasters, and especially none for authority figures, for example, plant officials, who pretended to know what they didn't. Laziness, fishermen who just sat around, people of negative attitude, those inclined to say, "Can't" were all shunned.

He was proud of his ability to get the most out of people, and he was a hard man to say no to when he asked a favour. But giving, in the Christian sense, was also important to him. He was known to help other men along. His wife, Ruth, put it this way:

> If he saw anything good in a man, he'd try to bring that out. He'd say, "Come on, son. You'll be skipper some day. But you've got to come up the way I did. You've got to learn to take orders. Because, by and by, you'll be giving them."

By one count, at least fourteen men who later became masters of their own trawlers had served under his command at one time or another. Arch's recommendation gave each of them an important boost up the ladder, sometimes despite their own self-doubt. By contrast, some masters had miserly reputations when it came to giving younger men opportunities.

Like his peers, he was keen to express his pride in the vessels and men he commanded. When speaking of his crew on the *Blue Foam*, he would declare, for example, that he carried the "pride of Newfoundland."

He would argue that he was not unique among skippers and other seafaring men for the qualities mentioned above. Other skippers of his generation were more or less like him in these respects. In fairness, however, not every skipper's opinion carries equal weight, and Arch's came to be highly influential.

Eagerness and competition on the banks caused him to engage in the same fishing tricks used by his peers, as in, for example, manipulating information shared over the *Blue Foam*'s radio. He'd say, "Those are only white lies. There's no harm in that." Nevertheless, he was a highly moral individual, and very conscious of family responsibility. He would take a young orphan "off the turf" for his first chance to the

The *Blue Foam* steaming from the Grand Banks with about 200,000 lbs. of fresh fish aboard. Photo taken by captain of the *Fairtry*, English factory trawler, c.1954

Epilogue: The Voyage Ended

banks when many more experienced men were available, just to give the lad an opportunity to help his family. But he had little patience for men who drank to excess, for in his experience, doing so risked lives and family well-being. That he never drank, never lost a man, and took a remarkable quantity of fish in his long career, seemed to confirm the wisdom of his moral approach and he was proud to say so. But, he stressed to me, some skippers who lost men "probably forgot more than I ever knew. I was just lucky. Because they had to do the same things I did." Modesty balanced his pride.

Skipper Arch's moral reputation makes him stand out. He was known as "smilin' Arch" in the communities along Newfoundland's Southern Shore, where banks schooners called for caplin bait each summer. Maggie Keough, age ninety-one in 1984, a Calvert, southern shore woman, remembered him as a loner, a fellow who kept to himself rather more than most banking skippers. Perhaps this was because, by all accounts, he neither drank nor smoked. Moreover, he was never know to blaspheme. His oaths were limited to such expressions as "Moses!", "Gee Whiz!", "By Jingoes!", and the rare scatological euphemism. He was like many Newfoundland fishermen in these respects.

Newfoundland and Canadian banking schooner fishermen followed the idea seriously that Sunday was the Sabbath, a day of rest. Arch never fished on Sundays until he began fresh fishing with the *Ronald George,* and, later, when trawling. He felt doing so unavoidable because the owners felt it necessary and fish might spoil otherwise. Although he expected forgiveness from an understanding Lord for this transgression, I am told that he declined to take communion at his church during his dragging years. But he was a regular churchgoer when in port. He loved most hymns and, at sea, often sang "Jesus, Saviour, Pilot Me" when he took his boat's wheel.

For all this, he was remembered as tolerant, not one to "wear his religion on his coat sleeve." Most fishermen of his generation were undemonstrative in this respect. A voyage could be half over before another's denomination was known. One's religion did not interfere with strong friendships with other men, like the eager Captain Clar' Williams, of more liberal habits. Perhaps what counted most among these men were seamanship, mutual trust, cooperativeness, eagerness, reliability, guts and performance, wit and humour.

The willingness to sacrifice and share the fruit of one's labours with one's family is writ large in his account of growing up and fishing as a young man. This agrees with his family's memories of him as husband and father. At settle-up time he always brought something extra home for his family. For example, he bought his father's good suit, the same suit he "carried away with him when he went."

His four children remembered him as a moral, supportive, compassionate, understanding and generous father. Their mother and father were totally dedicated to each other, and respected each other's different achievements. As children will, at first they took for granted what their parents provided, but they eventually became keenly aware of their sacrifices for them. As Catherine, the eldest daughter put it, "Dad just worked his fingers to the bone. He'd come home, wouldn't have enough money, and he'd have to go away and be gone for Christmas."

The children looked for him to be home each winter. At home, he'd warm their clothes by the stove at breakfast time before they left for school. If necessary to assert control over them, he did so with his voice and facial expression, and "never laid a hand on his children." When Roland faltered in his first effort to learn how to tie the bowline and other knots as a young Scout, Arch drilled him skilfully so that he quickly went to the head of his troop. Throughout his life it seemed that his greatest pleasure was in helping each of them along.

Like most youngsters in Grand Bank anxious to see their father return from long absences at sea, little Catherine and Florence would go to wait on the corner within sight of their

Epilogue: The Voyage Ended

Skipper Arch and son, Cyril, aboard the *Blue Foam*, in Grand Bank harbour, May 1948.

mother at home. As soon as Arch hove into sight they raced to meet him. Then he'd take them to visit the ice cream parlour in town or, a special treat where autos were still a rarity, for a taxi ride to and from the parlour in the nearby town of Fortune.

In the early morning when he left to go banks fishing before his children rose, he always left a 20¢ "shilling" piece under each of their egg cups on the breakfast table.

Early in his story we learned how the people of Anderson's Cove gave up their Anglican denominational affiliation for the Congregational one — in order to improve their children's educational chances. Arch and Ruth were similarly dedicated to their children's education. For example, a piano and musical training for the girls were major concerns for both parents. They did all they could to encourage them to get as much education as possible.

Despite his personal success and dramatic improvement in his family's income and well-being when on the *Blue Foam*, Arch was always bothered by his lack of formal education. If he could prevent it, his sons would not have to go to sea for a living. The hardships were made clear to both Roland and Cyril. In their teen years both took many trips to the banks with their father to learn the life at first hand. And they concluded that they would not be able to do what their father did — and **had** to do. Neither son followed the sea. For Catherine, finding a man to equal her father seemed a very remote possibility. At first she thought it would have to be a man who didn't follow the sea. In the end, she married a promising career naval officer.

In a deep sea fishing and seafaring community, it is understood that the sea will take its toll in lives. How, then, did the captain's frequent absences affect his family? On one hand, his children feel the circumstance brought the family closer together. And, it seems, girded by their mother's religious faith and example, they didn't worry while she often did. When Arch and the *J.E.Conrad* were missing, Florence recalled that she never ever thought her dad was gone or never ever coming home. She remembers no sense of anxiety. His absence was taken as natural. But when the family moved to St. John's, and he was home one night in nine for the first time, "it was a big deal!" Only ten years later, on 9 February 1958, the trawler *Blue Wave*, based in Grand Bank, was lost with all hands. This event introduced an element of fear in what his children had come to accept as a "natural" way of life.

The *Blue Wave* carried sixteen hands to their death. Arch was deeply shaken, as were other fishermen familiar with the boat and its crew. Fourteen of the sixteen men lost had sailed with Arch either on the schooner *Ronald George*, or on the *Blue Foam*, for from one to twelve years. As Arch told me,

> The bosun and mate, first, second and third engineers — all officers but the captain, and the cook and the rest, everyone sailed with me on the *Blue Foam*. And they were lost. It was just like a family. And it just about got me. Two men in particular had slept in the same berths aboard the *Ronald George* for four years, and eight years in the *Blue Foam*. 'Twas pretty sad.

He used to lay awake at night and call out their names.

The *Blue Wave*'s crew were mostly young married men. They left fifty-four dependents. Despite the wide-spread industrial and social assistance developments, and growing government intervention that accompanied Newfoundland's confederation with Canada in 1949, provisions for essential marine disaster financial relief had changed little since the turn of the century. Moreover, trawlermen were still considered "co-adventurers" rather than "employees," as they have been since 1971. Thus the surviving families were not entitled to aid under Canada's "Workmen's Compensation Act." Yet few hands, if any, carried life insurance. As in years past, the dependents' lot fell largely to the mercy of their family, friends, neighbours and churches. Fortunately, local leaders undertook a province-wide fund drive in their behalf, one similar to that undertaken in the wake of the

Ocean Ranger oil rig disaster in February 1982. The sum resulting from the *Blue Wave* drive was put in trust to assure yearly and monthly payments to the dependents.

The *Blue Wave* loss brought controversy over the design and operation of the same fishing machines that had brought new employment and prosperity to a growing number of fishing centres. Were they as safe as once thought? Under what conditions? The Canadian Department of Transport's Board of Steamship Inspection inquiry into the disaster seemed to find no fault. By their assessment, the vessel iced up and capsized. There was no suggestion that she was unstable to begin with, as many who had sailed aboard her later said they had come to believe. In the context of scarce employment opportunities and masters prideful of their commands and reputations, it seems such criticism and doubts were held back. Many trawler fishermen and families, however, found confirmation of their suspicions in other boat and crew losses, especially that of 18 February 1966, when the *Blue Mist II*, another trawler in Grand Bank's Bonavista Cold Storage Company fleet, was lost with all thirteen hands. Once again, the Board of Steamship Inspection inquiry that followed found no fault with the boat. But many trawler fishermen, including skipper Arch, were unconvinced.

Within months of the *Blue Wave*'s loss, Skipper Arch had a heart attack. His doctor told him to take things easier, to consider not going to sea again. Following a short rest, however, he was at sea again for several months before being struck by a second attack just as he was going through the Narrows of St. John's harbour, bound for the Grand Banks. The situation was serious. No pension, only what he had managed to save from his fishing. With little formal education and only his long fishing and command experience, like other fishing skippers with his background, there seemed few attractive alternatives ashore. He talked of looking for a shore job as night watchman.

Fortune turned his way once more, however, in 1961,

when he landed a job as skipper of the *Makkovik*, a Department of Public Works motor vessel based in St. John's and used in surveys and other coastal work. With a steady income, released from the strain of fishing and able to be in port frequently, the following years proved more calm until he reached retirement age in the mid 1960s.

But he was restless. Recognizing his wide knowledge and human relations skills, the Newfoundland Fisheries Development Authority recruited him as a "trouble shooter," to "feel out" fishery development projects around the province. Karl Sullivan, who worked with him then, recalled he was not reckoned a man for the office (he couldn't sit still). Arch took to the fact-finding trips with gusto, as he exchanged ideas with fishermen of all kinds.

The sea was always on his mind. He would always head straight to the dock if there was a large vessel in port. Then straight to the cook for a cup of coffee. "I'd like to be sitting on those benches. Every boat brings back a memory."

Not long before he died, he told his daughter Florence in Montreal, "One of these days I'll be going out again in the *Blue Foam* or one like her." But with the same reluctance to turn from the sea, when going to the hospital the morning he died, Arch told the doctor, "I don't think I'll be any good to go out on the Grand Banks tomorrow."

He left the way the Rev. Hugh J. MacDermott (1938: 177) described the last moments of another steadfast banking fisherman,

> ... with the same simplicity and fearlessness he went out to his last cruise, where no rough seas, no hardships, and no uncertainty of fog could dim the clear skies nor hide the face of his Captain.

The old man wanted no mourning at the bar when he put out to sea. It was a time to rejoice. His voyage was ended.

Retired banks fishermen pose in the sun. At dockside in Grand Bank harbour, c.1960.

Endnotes

Preface

1 Four less accessible and more academic essays published between 1978 and 1988 reflect a continuing effort to understand and explain aspects of the banks dory fishery organization and experience. Their ideas inform the introductory commentaries in Voyage to the Grand Banks. They are: "The 'count' and the 'share': offshore fishermen and changing incentives," in *Proceedings of the Canadian Ethnology Society meetings*, at Halifax, 1977. Edited by Richard Preston, pp. 27-43. Mercury Series, National Museum of Man, Ottawa, 1978; "Social organization of Newfoundland banking schooner cod fishery, circa. 1880-1948," in *Seamen in Society*, Proceedings of the International Commission for Maritime History meetings at Bucharest. Paul Adam, editor, pp. 69-81. Paris: 1980; "Recollections of Struggle." *Proceedings of the IVth International Oral History Congress* (1982), at Aix-en Provence, France. Francois Bedarida, et al, editor, pp.175-185; and "Usufruct and contradiction: territorial custom and abuse in Newfoundland's banks schooner and dory fishery." *(MAST) Maritime Anthropological Studies* (1988) Vol. 1, No. 2: 81-102. These essays are available at Memorial University's Center for Newfoundland Studies.

"Recollections of struggle" (1982) explores memory in oral history. It examines the reliability of Captain Thornhill's recall of events described in "Bound for Burin"(1977) — Chapter 6 in this volume. I interviewed the Mate, Garfield Rogers, and Ambrose Green, another crew member, who were present on the same futile journey. Both resided in Grand Bank. To no surprise, and for a variety of reasons, this comparison finds that individuals recall the same events somewhat differently.

Introduction

1 My interest in deep sea fishing focused on men's lives. While I listened for the landward part of their personal lives, and to and about their wives and families, that side required a focused study of its own. It seemed self-evident that the absence of husbands and fathers placed heavy responsibilities on mothers, children often hardly knew their fathers, and tensions and conflict were often high during the short periods fathers were home. A man lost at sea, of course, was a catastrophe, especially for widows and offspring.

Cynthia Boyd discusses some of these matters from the perspective of Grand Bank women in her interesting essay, "Come on, all the crowd, on the beach!", in *How deep is the ocean?* Edited by James E. Candow and Carol Corbin, pp.175-184. (Sydney, N.S.: University College of Cape Breton Press, 1997). H.C. Murray's, *More than fifty per cent: Women's life in a Newfoundland Outport, 1900-1950*, describes women's

lives in the inshore fishery. (St.John's: Breakwater Press, 1979). On the lives of women linked with modern trawler fishermen fishing from Nova Scotian ports, see Marian Binkley's, *Voices from off shore*. (St.John's: Memorial University, Inst. of Soc. and Econ. Research, 1994).
2 On the unionization of Newfoundland's inshore fishermen, fish plant workers, and offshore trawlermen, see Gordon Inglis, *More than just a union: The story of the NFFAWU*. St.John's: Jesperson Press, 1985. Among other important issues, Inglis describes how the "co-venture" system came to its end in 1974 (pp. 209-221.) See also David Macdonald's account of union leadership in the 1974-75 strike involving Newfoundland's unionized trawler fishermen: *Power begins at the cod end*. (St. John's: Memorial University, Inst. of Soc. & Econ. Res., 1980.)

Chapter 1 — An Excellent Place for Fishing

1 Garfield Fizzard's short biography, *Captain Frank Thornhill, Master of his craft* (St. John's: Grand Bank Heritage Society, 1988), provides interesting information on the Anderson's Cove and its people, and on a deep sea fishing career comparable to Captain Thornhill's account. Frank and Arch were cousins, sometimes shipmates, and competing skippers.
2 Long before the 1870s Long Harbour had a reputation as a "hunter's paradise," even among English and French naval officers and aristocrats. French fishing interests, of course, had rather exclusively controlled the south coast from the sixteenth century until the Treaty of Utrecht in 1713 (see Fitz-Gerald 1935, pp. 100-102, 118).
3 David Alexander (1980, pp. 11-12) estimated that, in the period 1836-1900, the south and west coasts of Newfoundland had the highest illiteracy rates relative to other regions of Newfoundland. "In 1891, at the very least, about 32% of Newfoundland's population over ten was totally illiterate." Further, "...there was no country responsible for its affairs and the progress of its people which drew upon such a meagre supply of educated people for its entrepreneurial, managerial and administrative requirements." Alexander reasoned that this deficiency limited the dominion's ability to effectively deal with the outside world (p. 48).
4 This action might suggest otherwise, but Congregational Church educators may have gone to extremes not to be denominationally competitive, having withdrawn from settlements in Notre Dame Bay and Trinity Bay in the face of the "aggressive denominational policy of others" (MacDermott 1938, p. 107). They sought to avoid settlements where other churches had already established schools, since overlapping created poorly equipped and staffed schools (pp. 78-79).
5 This was Dr. Conrad Fitz-Gerald, born in Marlborough, England, in 1847. To the Fortune Bay people he served, "the Doctor" was a hardy and intrepid sailing and hunting enthusiast whose exploits are still

recalled. This Marlborough doctor's son first apprenticed from age 16 to 19 with a registered medical doctor at Bath, before entering the Bristol School of Medicine in 1866. By 1870 he was a qualified surgeon and apothecary. Fascinated by the sea, he spent the next three years as a ship's doctor. In 1873 the London firm, Newman, Hunt and Co., recruited him to treat the firm's fisherman traders at Harbour Breton, on Newfoundland's south coast. This branch of "Newman's" supplied all the smaller stores in Fortune Bay. His practise included all of the Bay settlements. The firm's fortunes had faded by 1901, when Fitz-Gerald moved to nearby St. Jacques, where he continued his practise almost until his death in 1939 at ninety-two. Conrad Trelawney Fitz-Gerald, Jr., a grandson, published his biography, *The "Albatross"*, in 1935. The Doctor is also mentioned in Allen Evans', *The Splendour of St. Jacques*. (St. John's: Harry Cuff Publications, 1981, pp. 43-46). Evans suggests that The Doctor's yearly fee was $4.20 per family, and that "in many, many cases the poor souls couldn't even come up with that amount" (p. 43). Yet, his biographer reports, he doled out medicine at no charge for about 40 years (p. 164); probably with the help of an endowment given him by an aunt for this purpose (Evans 1981, p. 43). In 1877 it seems the fee was only $4.00, paid in fish value to Newman's, who in turn paid The Doctor in cash for, "Between merchants and fishermen there were no cash transactions" (Fitz-Gerald 1935, p. 105).

6 Arch Williams, former Wireless operator at Pool's Cove, Fortune Bay, opines that Captain Thornhill refers to the compound copper sulphate, a germicidal. But brimstone was the local term for sulphur, available in block or powder form from area merchants for traditional remedies used internally and externally. For example, Fortune Bay fishermen believed a spoon of powdered brimstone mixed with a glass of molasses (hence, the centuries-old prescription "brimstone and treacle") and drunk, helped purify one's blood. For some, it was a standard 'spring tonic' given children daily for a week to cleanse the blood and start the new year soundly. Such tonics were widely used in Newfoundland.

7 The *Mamie and Mona* may have been a 43 or 45 ton, four dory vessel built around 1904 for Captain J.M. Fudge of nearby Belleoram. In 1905, Captain Fudge used a vessel he called the *Mona and Minnie* on the Western Banks, Grand Bank, French Shore and Labrador (Fudge 1963, pp. 10-11). Later in this same memoir he mentions the *Mona and Memmie* (p. 35), in which he made a late fall run to Prince Edward island for produce in 1908. In ten days he sailed to St. Pierre, Prince Edward Island, and back to Belleoram with a full load of produce. The three names may refer to one and the same vessel.

8 Gloucester and Fortune Bay fishermen did not always enjoy friendly relations. In the late 19th century herring for bait was at the center of confrontations between Fortune Bay and American fishermen, and

Anglo-American diplomatic haggling. The pivotal incident occurred during the first week of January, 1878, when twenty-two American schooners from Massachusetts were gathered in Fortune Bay awaiting the herring. On Sunday, 6 January, boats from four of the American shooners began to seine herring in Long Harbour. The local inhabitants objected to the Americans fishing on the Sabbath, and they feared that their own catches of fish would diminish if the taking of herring continued. The Newfoundlanders forced the American fishermen to stop their seining operations and to release fish already caught. In one instance they destroyed an American seine after dragging it on shore. An American captain inflamed tempers further by threatening the local residents with his revolver. Consequently, the Massachusetts fishing vessels were forced to return home without a profitable catch of herring (Pennanen 1979, p. 289). The incident strained Anglo-American diplomatic and economic relations. Newfoundlanders were castigated for "acts of lawless violence" contrary to the Treaty of Washington of 1871, which appeared to give American fishermen the right to fish within three nautical miles of Newfoundland shores (pp. 289-290; and see A.M. Fraser's examination of these events, their resolution and wider legal implications within the context of fishery negotiations with the United States in the period 1783-1910, 1946, pp. 348-355).

Chapter 2 — I Never Gave Up Once in My Life

1 Stan, one of Skipper Arch's younger brothers, (b, 1904), recalled the *Sanuand* as a twenty ton, two dory schooner. It fished in the Straits and off St. Pierre, St. Mary's, and the Labrador. But not on the Grand Banks. It was probably a "jack boat," or "western boat," built on Newfoundland's south coast. It has a transom stern and the rudder "outdoors," to be free of ice in the winter time. (The alternative was the usual schooner with overhanging stern or counter, with the rudder stock up inside it through a trunk. In winter, the rudder post was liable to freeze up in the trunk, which also tended to develop leaks.) H.F. Pullen describes the western or jack boat:

> These transom-sterned vessels were generally between 45 and 50 feet long, fore-and-aft rigged with a long bowsprit. The sail plan was quite simple, consisting of two jibs, a foresail and a mainsail. Topmasts were not fitted. The hull was designed to produce a graceful, fast and seaworthy vessel, well suited to the rough and tempestuous conditions to be found in Newfoundland waters (Pullen 1967, p. 35).

2 One might think engine-power meant only gains for inshore fishermen, in time saved getting bait, fishing, and increased time ashore. But one non-fisherman observer of the Fortune Bay fishing scene from 1904 to 1934 opined that, too often, the added expense of capital and running costs were of little value to him, and he had nothing to do during his

spare hours ashore to help his (largely non-cash) income. Moreover, the catch was often too small to pay for running expenses and salt for curing. Further, "... small traders, and agents, in those early days of the marine motor, forced engines on many fishermen; and the fishermen only too eagerly "took up" engines with no prospect of paying for them" (MacDermott 1938, p. 180). The added debt, and inevitable breakdowns when the battery ran down and the magneto ceased to give the engine life, made for resentment against supplier and cleavage between fishermen and merchants.

3　The salmon sport fishermen and hunters attracted to the Long Harbour hunting grounds provided some seasonal employment to Bay d'Espoir Mi'kmaq Indians. Captain Thornhill recalled they often frequented the area and were "great guides." One, Steve Bernard, played the violin and often visited Anderson's Cove and Stone's Cove to play at dances at Christmas.

4　By the end of the nineteenth century, West Country English mercantile houses, like Newman, Hunt and Company operating out of Harbour Breton, which had monopolized Newfoundland's trade, had yielded their ground to parvenu domestic entrepreneurs and merchant houses concentrated on Water Street in the Newfoundland capital. Like their predecessors from England, these merchants dominated outport merchant trade as major sources of capital and merchandise. They supplied goods to fishermen and smaller outport dealers — in effect, their agents — on a credit basis, against local saltfish production. The St. John's merchants exported the fish, often in their own vessels. Doubtless, the local agent of such powerful patrons had more weight in local affairs than ordinary producers of fish. It is common to hear such agents characterized as the "king" or "uncrowned king" over a settlement (Wareham 1975, p. 17).

5　On a Sunday in 1904, a medical missionary, Rev. Dr. Hugh J. MacDermott, arrived at Belleoram aboard the S.S. *Restigouche* to begin 30 years of work on Newfoundland's south coast under the Congregational Colonial Missionary Society. He was met by Uncle Charlie in his trader, *Elsie*, and taken along the Fortune Bay coast, stopping here and there along the way, before landing at English Harbour East. MacDermott later settled in Pool's Cove (MacDermott 1938, p. 29). His recollections of 30 years service in the area reveal much about the man and the way of life in this area during these years.

6　Cyril Thornhill, another of Skipper Arch's younger brothers, ran Uncle Charlie's store for a few years, then he and Arch bought Uncle Charlie's interest. Later, Cyril bought Arch out. Cyril eventually sold it to Leslie Pope, then moved to St. John's. There he started a boarding house near Arch's residence and worked as a clerk at Ayre's, a large local department store. He was last employed in St. John's by the Avalon School Board and is now retired.

7 The 'kedgie' was usually a young boy of 15 or younger out on his first trip to the banks. His primary duties were on deck where he was to catch dory painters as they came alongside to discharge their catch, wash the decks down, keep the fog horn going, and help the cook in the focs'le. In 1918, he received $20 for the voyage, and cut out cod tongues and cod sounds from which he made, if he worked hard, an extra $100. Skipper Arch recalls that most dory fishermen had a very good attitude toward the kedgie, treated him well and gave him helpful advice when he needed it. Should a dory fisherman be ill, the kedgie often went out in his place and received a share in the catch. But most of his time was spent on deck, working sixteen hour days most of the time, Sundays excepted. He also recalled that the French schooners, which often fished beside them on the same grounds, called them "galley-boys," and treated some barbarously.

8 It seems there was a standing rivalry between Captain John Barnes, and his *Thistle*, and Captain Joe Bullen of Bay de L'eau, with his "almost identical" *Dixie*, over which of these four-dory, cutwater bow vessels was the faster. Both may have been 40 ton vessels (Evans 1981, p. 26).

9 South Coast Newfoundland fishermen, and merchants, once enjoyed a lucrative trade with St. Pierre, selling herring and caplin for bait. In 1869, for example, Newfoundland fishermen earned an estimated £27 to 28,000 Sterling for herring, and £16 to 18,000 Sterling for caplin; a total of about £45,000 (Newfoundland, *Journal of the House of Assembly* 1869, p. 512).

10 Article 5 of the Treaty of Versailles in 1783, assured French fishing interests the right to catch fish and land and dry it on the Newfoundland shore, from "Cape St. John, passing to the north, and descending by the western coast of the island...to Cape Ray" (Rowe 1980, p. 314). This was a small part of the coastline controlled by the French in earlier years. The French shore area Captain Thornhill refers to here is chiefly the stretch from Cape Bauld, south to Cape St. John (Nauss Cluett, pers. comm.. See also Fraser 1947, pp. 275-332).

11 'Banquero' is also referred to subsequently as 'Banquereau,' 'Quero Bank,' and 'Quero.' They are one and the same.

12 This is probably the Congregational Church doctor, Rev. Hugh J. MacDermott, mentioned earlier.

13 The Doctor and older people had reason to recognize diptheria's symptoms and dangers, and reason for caution in publically pronouncing its presence. It had raged in the Fortune Bay area for months in 1888, sometimes near panic proportions. Funerals were an everyday occurrence, and one was particularly gruesome. It consisted simply of two men carrying a coffin which was slung on a long pole. The people were

evidently near panic, and it was almost impossible to get help to place the dead in coffins. In a small dwelling in Belleoram the doctor found a whole family of six dying. The mother and father lay on the floor of the kitchen in a semi-conscious state, while four small children gasped for breath in an adjacent room. Antitoxin was unknown at this time, and it was almost impossible to cope with the disease. Carbolic acid was used extensively as an antiseptic, but often resulted in bad burns (Fitz-Gerald 1935, p. 168).

14 Twenty dollars may have been a standard sum paid the experienced "dory boss" or "dory skipper" at this time. In July 1980, Captain Jim Harris (b. 1894 at St. Joseph's), another retired south coast Banks dory fishing schooner skipper, recalled that in 1910, on the eve of a voyage aboard the four dory, 49 ton schooner, *Maggie Dunford*, he suddenly found himself without a dorymate. V.T. Cheeseman, a Port au Bras firm, paid him $160 as dory boss plus $20 to take a man without banks fishing experience. "This extra $20 was deducted from the dorymate's wages, so he got $120." But in 1976, Skipper Arch knew no other case of anyone having bought a chance to go in dory as he did. Arch's younger brother, Stan (b. 1904), argues that Arch had to do this, because he had no relative to take him in dory, especially once his older brother, Jim, entered military service.

15 A short account of Captain Thornhill's first opportunity as a banks dory fisherman is given in an article in the St. John's *Free Press*, September 1, 1971. It includes details on his career and family.

Chapter 3 — Voyage to the Grand Banks

1 See Garfield Fizzard's, *Unto the sea: A history of Grand Bank*. (St. John's: Grand Bank Heritage Society, 1987).

2 I found no official record of the *Bessie MacDonald* and her specifications. In October 1977, however, retired banks dory fisherman, Mr. Garfield ("Gar") Rogers (b. 1900), of Grand Bank, informed me that he had once sailed on her. He described the vessel as an eight-dory, cutwater schooner of about 90 tons, about 90 feet in length and 22 feet in beam. Further, she was built at Grand Bank about 1908 by merchant Eli Harris, and named after the local doctor's daughter. She was called the "Wooden Wall" by those who sailed in her, because she would hardly list when under full sail. Mr. Rogers believes she was sold in 1923 by Samuel Harris, along with his other schooners when his Grand Bank firm went "broke."

3 A 'sed,' 'sud line,' or 'snood' line, also variously 'gangin',' and, elsewhere, "gangling" (Leather 1970, p. 252), is a line about three feet long, attached at varying intervals — depending on ground conditions — to the main or trawl line. A baited hook is attached to each sed.

4 A shipping paper, as illustrated here, bound each man to his vessel. This agreement committed the man to work for the vessel owner(s)

from the outset of the voyage in the spring until its conclusion on an unspecified date, usually in October. The shipping agreement given here is a facsimile based upon an actual one dated 1912 drawn between William Forsey, another Grand Bank vessel merchant, and another man. Similar documents were used by other Grand Bank vessel owners at this time. It seems that the document was validated by the merchant's signature, whether witnessed or not. And did not seem to require the crew member's signature at all, at least not in this case. How the local magistrate, or any other Newfoundland court, would see the situation is uncertain.

5   The French Islands' proximity to Newfoundland, especially its south coast, was either a boon or devilish matter, depending upon where one stood on the use of alcohol. Captain Thornhill does not elaborate on the controversy during his years on the south coast, but they were marked by an old struggle between "rum and religion." F.W. Rowe (1980, p. 257) writes:

> The temperance and total abstinence movement became so powerful that, by 1915, a majority of the people were in favour of complete prohibition. As in other countries, prohibition soon proved to be unworkable. Smuggling and moonshining were rife, and the provision whereby medical doctors could issue prescriptions for alcohol soon became open to widespread abuse. Prohibition was repealed in 1924.

Captain Thornhill was among those of his generation who favoured and practised total abstinence.

6   The lives of Newfoundland's fishermen give countless illustrations, and lessons, in the power of unyielding will to survive and succeed. The Rev. Dr. Hugh MacDermott (1938, pp. 265-266) tells of one such man living not far from Belleoram:

> He had had a hard life, and seemed tired out, glad to rest now that his family were grown up. On one trip he had been astray in his dory from his vessel, just as has happened to many others. For days he had drifted in the Atlantic. Driven to desperation with thirst, he had chopped his fingers with his splitting knife, and sucked them. He lived and was rescued. One of the small and hardy type of men, he continued to fish at the Banks later in life than most men.

And young men were doubtless driven by the teachings of others, whose lives were testimony to the truth of their forceful moral dictums. For example, Captain J.M. Fudge (1963, p. 6) of Belleoram, who also suffered the ravages of sea sickness in his first trip to the Banks, wrote, "...life is a battle and we one and all need good clean courage, moral and spiritual, to win out. For a quitter never wins." And for men like Captain Thornhill, who took their Christian rearing seriously, given

pure intent, their willful labours were strengthened by faith that a watchful and benevolent Lord will provide.
7 Skipper Arch could not recall ever going to a dance in Grand Bank before his marriage in 1928.
8 The older women of these caplin ports on Newfoundland's 'Southern Shore' especially remember the excitement of planning parties and dances, and dressing up when the bankers arrived and their young fishermen came ashore. Their arrival marked the close of a long winter of relative isolation. Further north, on the east coast, where many local men were away to the Labrador fishery for the summer, banks fishermen had more room to socialize when making port. For example, Andrew Horwood (1971, p. 164), recalling his experience as a youth at Carbonear, where bankers seeking squid for bait often crowded the harbour, wrote:

> Usually the men who were called bankers would have very little money to buy anything other than squid but there were small shops that sold home-make beer and confectionery. The owners looked forward to the arrival of the bankers. Some of the boys from the vessels found the girls that they would like to have for wives. They never did take one away in the ship but the girls were given instructions that enabled them to find their men in the fall when the fishing season would be over. And it worked out. The young men of Carbonear who were away on Labrador while the bankers were in, regarded those matches with little sympathy.

9 William Wallace (1955, pp. 333-391) relates that the first such German submarine raid occurred in October, 1916. There were none in 1917. Six submarines resumed their trans-Atlantic raids between April and October 1918. The Canadian Navy responded with regular navy patrols and six armed decoy schooners of 94-100 tons. Each schooner carried a navy gun crew and 12 pounder, a 12 cwt rapid fire cannon mounted on deck aft of the foremast, and disguised by a screen of false dories. They were also fitted with 30 h.p. diesel engines. Wallace was sailing master of one of these schooners, the *Albert J. Lutz*, in 1917.
10 According to Wilson J. Osborne, late marine blacksmith at Grand Bank, this was the whaler *Hurricane*.
11 In August 1918, German *U-boat 156* captured the Canadian steam trawler *Triumph*. With a prize crew onboard, she became an armed raider hunting on the fishing grounds off Nova Scotia. She approached and sank four fishing schooners. Their crews were allowed to take to their dories in each case. Later the Germans sank the trawler with time bombs. Returning home to Germany that October, *U-boat 156* struck a mine in the North Sea and sank with all hands (Thomas 1973, p. 166).
12 'Settle up' time sometimes meant more than the resolution of individ-

ual financial accounts with the merchant. It was also often the time to settle personal grudges developed during the voyage but deferred until its conclusion. Linking the settlement of personal disputes with the fishing voyage calendar may have a long history on the south coast. Some writers suggest such disputes approximated annual rituals. At Harbour Breton during the mid- to late-nineteenth century, settle up "day" was a week-long affair that began regularly on October 10, and assumed the character of an annual combative, grudge settlement event, "...the champion for the year being the chap who had defeated all challengers" (Evans 1981, p. 28-29). And the combat kept the local doctor busy (Fitz-Gerald 1935, pp. 105-107). Cheap rum from nearby St. Pierre and Miquelon fueled the settle-up celebrations and disorder (Bown, 1971, pp. 35-36).

13 See Otto Kelland's *Dories and dorymen* (St.John's: Robinson-Blackmore Press, 1984.) account of the development of dories and fishermen experiences with them.

14 In the 1890s and early years of this century, the widow or other dependent(s) of a fisherman received a "government grant" of $80 drawn from the "Bank Fishermen's Insurance Fund." The fisherman, or the merchant on his behalf, contributed 50 cents to this purpose each spring when he signed on. Additional voluntary assistance came from one's local church, and from the "permanent Marine Disaster Committee of St. John's" (MacDermott 1938, p. 189).

15 Captain Smith and his crew were evidently fishing on "average shares" on this occasion, rather than counting fish. These incentive methods are compared in greater detail in the following chapter.

16 Captain Clar (Clarence) Williams, an outstanding Fortune Bay banking schooner fishing and foreign-going master, stressed that the captain's main concern was to have good splitters. With them, a crew could salt their fish in "half the time it usually took" (Chard 1974). Beyond the speed factor, good splitters produced better quality split fish.

17 These helpful, and sometimes dangerous, deck or 'hoisting' engines probably began to appear on vessels operating from Grand Bank prior to 1915 (Smith 1974). Likewise, auxiliary power began to appear in Gloucester schooners by 1912. Captain Sam Bartlett of Brigus, Newfoundland, reportedly had a topsail schooner, *Laddie*, of about 100 tons with auxiliary power in 1912 (N. Smith 1936, p. 12). But fully rigged schooners were still being built in New England until the 1920s (Leather 1970, pp. 257-259).

18 There seems to be no documentary record of what the *Flora S. Nickerson* landed in 1921. But the disappointment of a low catch was probably compounded by other factors hostile to Newfoundland's fisheries at this time. Some are specified in his 1921 annual report by W.F. Coaker, then Minister of Marine and Fisheries:

The fluctuations of exchange have had a very serious effect in every market. In every country the aftermath of the Great War has left its trail. The decreased purchasing power of the nations, heavy and increased taxation, competition compelled by motives similar to those governing our own, social unrest with accompanying disorders, strikes and their evils on the one hand taken together with large steamer shipments, anxiety to realize and consequent congestion of cargoes, high cost of production and taxes on our side have all made the time of realization a most anxious one for Newfoundland (*Annual Report of the Department of Marine and Fisheries* for 1921 (1922), p. 8.)

But many people blamed Coaker for the severe slump in the fish market. He was replaced as Minister in 1923. Coaker's removal did not alter the imperative need to totally reorganize the Dominion's fishing industry, and, particularly in the area of market development, to fully realize its resource advantages and sell its products in a changing and increasingly competitive international saltfish marketplace (Alexander 1977, p. 7). The industry, and Newfoundland's maritime economy generally, continued to wither through the '20s, Great Depression, Second World War, and following years. Even union with Canada failed to restore Newfoundland's potential "as a major world producer of fishery products." A global economic perspective on the industrial context in which our captain's experience occurs is provided by the late economic historian, David Alexander, in his *The decay of trade* (1977). It deals especially with Newfoundland's saltfish trade in the period 1935-1965.

19 Data in the *Annual Report of the Department of Marine and Fisheries* for 1922 (p. 15) suggest that an "average voyage" among the 20 vessels listed for the Grand Bank community fleet in 1922 was about 3,035 quintals. These vessels, averaging about 85 tons, landed 60,692 quintals that year.

20 J. Wilson Osborne, late Grand Bank marine blacksmith, opined that Skipper Arch's brother, Stanley, was in dory with their oldest brother, Jim, on this same trip.

21 A well-thumbed copy of *Fast and able*, Gordon W. Thomas's compilation of "life histories of great Gloucester fishing vessels" (Gloucester, MA: Gloucester 350th Anniversary Celebration, Inc. 1973), was among Captain Thornhill's books at his death. His description of the death of the *Rex* conforms closely with that recorded by Thomas (pp. 144-146).

22 In the confines of a small ship at sea a contagious disease may be catastrophic. The tragic sea ballad, "Bound down for Newfoundland" (Peacock 1965, Vol. 3, pp. 905-906) recalls one such incident involving smallpox. Only two of the schooner *Mary Ann*'s young crew of eight, bound from New York to Newfoundland, survived.

23 Captain John T. Thornhill was a banks skipperman of heroic proportions, noted for his aggressiveness and success when fishing, and for his "big ideas." Older Fortune Bay banks fishermen regarded him as "the grandfather" of the Newfoundland banks fishery, and believed he landed more fish than any other skipper in his day. He was always a highliner, "Lord, not only voyages, but on trips. If he just went outside of Grand Bank Cape, he could get a load of fish." When his vessel hove in sight of Grand Bank, you'd hear people say, "Here comes Skipper John, with a waterhorse (roughly, a large stack of fish) in tow." He built himself one of Grand Bank's largest homes, and had one of its first automobiles in about 1925. He died in the early 1940s. His home became the Thorndyke Hotel in the 1960s. Today it is a popular bed and breakfast.

Chapter 4 — My First Command

1 This was Ruth Williams (b. 1901). Her education at Berea College, Kentucky, was supported by a Grenfell bursary and assistance from the Congregational Board of Education under the Newfoundland Outport Nursing and Industrial Association (NONIA). After completing the Berea program she taught at the Day School, Little Bay East, where she also conducted Sunday School, Sunday Services, and preached on Sunday evenings (MacDermott 1938, p. 273).

2 Their exact frequency is unknown, but such mishaps were common to fishermen — American, Newfoundlander, Nova Scotian, and others — fishing these often fog-shrouded grounds. And they involved skippers and other crew members. For example, another instance is recorded in a dispatch dated July 14, 1894, from the U.S. Consul in St. John's to the U.S. Secretary of State in Washington, D.C. It reads:

Sir,
I have the honor to inform you of the arrival at this Consulate on the 12th inst. of Edward Muse of Schr. "Eleazor Boynton" Burke Master, Wm. parsons 2d & Co. owners of Gloucester, Mass., who had his hand shot off whilst discharging fog guns on the Grand Banks. The Captain deposited $25.00 at this Consulate for his benefit and when sufficiently recovered will forward him to the Consul General at Halifax."

3 This may have been Captain Bill Miles, who skippered a schooner owned by the English Harbour West merchant, Jerry Pettitte. Allen Evans (1981, pp. 11) reports that Captain Miles was left totally blind when he discharged the fog gun, or 'swivel,' and powder struck him in the face.

4 Bait supply was always a source of uncertainty for production in the banks salt fishery. One observer in 1889 estimated that a banker might lose four weeks of its yearly voyage merely seeking bait. This problem led to creation of a "Bait Intelligence Service" in 1890. It linked tele-

graph stations in 56 communities around the island to provide daily reports on bait location and abundance. Bankers eagerly used the service (*Newfoundland Fisheries Commission Annual Report* for 1889, pp. 45-46, and for 1891, p. 33). But the bait problem was not wholly eliminated.

In 1914, on a visit to Newfoundland, Walter Duff, of the Fishery Board of Scotland, was struck by the bait scarcity problem in July and August, when weather was ideal for fishing. He wrote:

> In Placentia Bay during the last two weeks of August it was pitiable indeed to observe a number of schooners, each with a crew of 20 to 25 men, waiting anxiously the arrival of the squid and making ineffectual attempts to secure small quantities of lance or herring. Not only the crews of these schooners but the shore fishermen were held up for the same reason. By the end of August the patience of many of them became exhausted, and I was informed that several of the schooner Captains discharged their crews rather than keep them doing nothing any longer. Now why should this be the case? Why should fishermen be lying idle for weeks for want of squid when other bait can be easily secured at a time when cod are in abundance almost at their very door...I am convinced that if fishermen were possessed of a few (herrring) drift nets a sufficient supply could be secured daily, to enable them to proceed with their work (1914, pp. 3-4).

The idea of using such nets had been advanced before. But to no effect.

5 In his otherwise informed account of "Life on the Grand Banks" (*National Geographic* July 1921, Vol. 40, No. 1, pp. 1-27), Frederick William Wallace compared the "market Banker" serving the New England fresh fish market to those engaged in salt banking, and reckoned that "...the salt fishermen, as a rule, take life easier" (p. 28)! Captain Thornhill's recollections challenge this view.

6 Characterizations of outport merchants in pre-Confederation (1949) Newfoundland as roguish, oppressive exploiters of fishermen and other working people are widely met. For example, a European nurse who served at Rencontre West, on the south coast, in 1934-40 found it difficult to comprehend the outport merchant's power over the people, especially those on the dole. Referring to the local merchant, she observed that he,

> demands that the people who are on dole sign blank receipts for the goods they receive and which the merchant distributes. The merchant then fills out these blanks at a later date! For years there has been some talk that...he was cheating them. But such talk was only behind the merchant's back..., and the people continued to sign the blanks, like obedient children. Once long

ago, the son-in-law of my landlady dared to challenge our merchant. Also someone from New Harbour, which is also served by our Rencontre West merchant, did the same. There was an investigation, but the result of it was...that they took away from the merchant the administration of the dole and handed it over to his son, who worked as his employee! I also heard...that some check frauds were detected (Dickman 1981, pp. 46-47).

But several themes vie in remembered relations with outport merchants. At other times they are protective, paternalistic figures working to help an ill-educated and dependent labour force face destructive natural and economic forces, and pillars of positive action in their communities, even when it seems against their financial interests. Thus one hears of merchants who burned their credit accounts or "threw them into the harbour," clearing the slate for a client hopelessly in debt to them. There are "redistributive" and self-maintaining functions in such actions, of course. He supplied the fishermen with necessary equipment, food and clothing in the spring of the year on credit. In the fall of the year, at the end of the fishing season, he purchased the fish. Seldom, however, was there any cash left over after the fisherman paid his bill. High prices were charged for supplies and low prices were paid for fish. But there were always some who never caught enough fish to pay their bills. This loss was covered by the gain on the "good fisherman" who was in effect paying for the "bad fisherman" who was also supplied by the merchant (Wareham 1975, p. 16). Further examination of such seemingly contradictory themes in the oral history of worker-fisherman and merchant relations belongs elsewhere. It is enough for us to acknowledge that differences between the two in power, life-style and advantage, were such that notions of a "feudal" social order are often applied to Newfoundland, then and now. (cf. Sider 1986; and Sweeney 1997.)

7 Skipper Arch had the only shares (2) owned in Newfoundland. The rest were owned in Lunenburg.

8 Their wedding took place at Pool's Cove. The Rev. Hugh J. MacDermott officiated. Everyone in town came to the reception at the bride's family home. Formal invitations were neither customary nor necessary. The newly married couple went to Anderson's Cove for the winter on December 4. Ruth, the bride, remembered their arrival: "We went down on the old *Glencoe*. The men were all lined up on the road, firing their guns. They nearly blew my head off." When told it was the bride and groom going ashore, the *Glencoe*'s Captain Blandford blew the ship's whistle. Another reception was held at the school. When the tables were cleared away, there was a dance accompanied by accordian music and attended by the entire community.

9 The *Annual Report of the Department of Marine and Fisheries* for 1929 (p. 23) indicates that Captain John T. Thornhill and the *Paloma* landed some 4,177 quintals that year. Only one other Newfoundland bank fishing vessel, the *Christie and Eleanor*, landed more (4,484 quintals) that year.

10 *Clara B. Creaser* was later renamed *Robert Max*, after Captain John T. Thornhill's two sons.

11 The tidal wave created by the earthquake offshore on November 18, 1929, struck the southeast part of the Burin Peninsula from around Burin to Lamaline and Allen's Island. It caused 27 deaths and an estimated two and one-half million dollars of property damage. Important marine ecological effects inshore noticed the following year may have embraced the area from Cape Race, in the east, to Cape Ray in the west. The *Annual Report of the Department of Marine and Fisheries* for 1930 (1931, pp. 8-9) held that the fishery in this area was "...almost a blank and the scarcity of bait has been unparalleled in the history of the country."

12 Allen Evans (1981, p. 26) recollects going as a young lad to Mose Ambrose, near St. Jacques, about this time to work as a lumper, discharging and washing out a vessel's catch, for 15¢ per hour.

13 The *Annual Report of the Department of Marine and Fisheries* for 1930 (p. 23) records this as 2,884 quintals; the *Robert Max* had the top catch among the five vessels fishing for the S. Harris & Co. firm that year. However, it seems the big fish killer among Newfoundland banks skippers that year was Captain George Follett in the *Freda M.*, with 4,165 quintals landed for G. & A. Buffett, another Grand Bank firm.

14 The schooner *Robert Max*, loaded with saltfish and bound for Lisbon, was sunk by German submarine shellfire early in World War II. Her captain, Harry A. Thomasen, and five man foreign-going crew were permitted time to escape in their lifeboat, and saw their vessel sink. They were three hundred miles from the nearest landfall. Captain Thomasen set their course for St. Michael's in the Azores. Rowing and sailing, they reached land safely in three days (Horwood 1973, pp. 79-83). Others were less fortunate.

15 The *Annual Report of the Department of Marine and Fisheries* for 1932 (p. 15) recorded 3,066 quintals for the *Robert Max*. It ranked an acceptable third among the S. Harris Co., Grand Bank, fleet of five schooners, whose catch ranged from 1,985 to 4,000 quintals.

16 Newfoundland's economy and workers at this time were in a desperate state. Some idea of the situation is given in the assessment of Lord Amulree, in his *Newfoundland Royal Commission Report* of 1933 (pp. 83-84). Fish prices began to fall in 1930 and reached their lowest level in the century by 1932. By the end of the 1932 season the average fisherman, according to Amulree, was hopelessly in debt to the merchant,

and had been reduced to abject poverty. During the winter of 1932, no less than 70,000 persons or 25 p.c. of the population were in receipt of public relief, other than poor relief or relief for the aged poor. Such relief was distributed in kind, i.e. in rations of pork, flour, tea, and molasses of the maximum value of $1.80 per head per month. In the winter of 1932-33, privation was general, clothing could not be replenished, credit was restricted, and hardly anywhere did the standard rise above a bare subsistence. Lack of nourishing food was undermining their health and stamina; cases of beri-beri...and of malnutrition were gradually increasing, and were to be found in numerous settlements; the general attitude of the people was one of bewilderment and hopelessness.

Indeed, the experience of these years scarred and wizened a generation in ways difficult to summarize here. The Depression experience, including high unemployment and greatly diminished opportunities to escape through emigration to Canada or the United States, heightened a readiness to break with a disillusioning past linked with the old saltfish industry. This readiness was reinforced by cash labour opportunities that came with World War II construction of American military bases in Newfoundland and Labrador, encouraged by new employment prospects in offshore trawling oriented to an emerging American fresh frozen fish market in the 1940s, and union with the larger Canadian economy in 1949. The mainland North American industrial way of life beckoned.

Chapter 5 — **Barbados She's Goin'?**

1 Newfoundland banks fishing skippers commonly navigated by compass, log, dead-reckoning, and their shipmate's and own experience. Many had little, if any, literacy. The valuable information on sea charts was of limited use to many. Fishing and foreign-going Captain Tom Bartlett of Grand Bank, informs me that many who 'went foreign,' with cargoes to and from Europe, as far east as Italy, and to the West Indies and Brazil, relied on the quadrant (and, likely, divider and parallel rule) which, unlike the sextant, enabled them to get only latitude positions. The skipper held an easterly course, at, say Lat. 40°N. for Oporto, for so many days, perhaps logging the distance covered, until making a landfall. Then, like navigators and seamen centuries before, "Should he make the land to one side or other of his port and be uncertain of his position he could usually land in what was always a civilized country and ask the way!" A story is told of one such skipper who was asked how he managed this and replied, "I stopped everything I saw on the ocean and asked the way." Experience and boldness may have enabled some to go foreign successfully even without either

quadrant or sextant. Some Newfoundland skippers and their crews were lost, of course.

The biography of Captain Harry Thomasen (Horwood 1973), another foreign-going skipper from Grand Bank, represents those who rather specialized in Newfoundland's shipping trade. Captain Thomasen, a native Dane, received his navigational training while a deckhand on a Danish steamship. He literally jumped ship and swam ashore at St. Lawrence, on the Burin Peninsula, as a young man.

Mr. Curt Forsey (b. 1898), late son and heir of a Grand Bank schooner fleet merchant, recalled that the scarcity of local skippers with adequate foreign-going navigational training was especially acute during the period circa 1910-1920, when the Grand Bank fleet was rapidly expanding. As a result, a skilled navigator, one Captain Levi Angus McCush, presumably a Scot, frequently came over from his base at Belleoram to train the local skippers.

2 Captain Clarence Williams, born at Pool's Cove in 1898 (d. 1976), was a highliner banks skipper and foreign-going master. As a young man, he went to the banks in the summer months and in winter on sailing vessels carrying salt cod fish to markets in Europe, the West Indies and Brazil. A strapping man, more than six feet tall, he took his first command at age twenty. He went to sea more than thirty years and skippered such schooners as the *Lillian M. Richards*, *Pauline C. Winters*, *L.A. Dunton* (now fitted out at the Maritime Museum in Mystic, Connecticut), and the three-masters *General Byng*, *General Trenchard*, and *General Currie*, owned by Samuel Harris Ltd. Ashore in the early 1950s, he contributed importantly as a member of the Walsh (1953) federal-provincial Commission on Newfoundland's fisheries, consultant to the Government, and as a teacher at the Newfoundland College of Fisheries. The Memorial University of Newfoundland awarded him an honorary Doctor of Laws degree in 1975. He and Skipper Arch remained fast friends throughout their lives, and their graves are on a hill with an excellent view of the sea through the Narrows of St. John's harbour.

3 Discarding the tally board signalled a major change in work conditions and relationships. The act symbolized departure from oppression and inequity toward more democratic, equitable crew relationships. The vessel owner or merchant seems to have played little, if any, role in this decision. It was determined between skippers and crew. But many especially successful banks fishermen saw the count as consistent with their basic work values and only right; the individual should receive rewards in proportion to his production. And, indeed, some banks fishermen, including skippers of Captain Thornhill's generation, have told me that the count was never completely discarded, not even in the darkest days of the Depression. And they express this view pridefully. Dealing with fragments of a rapidly retreating past, many questions

endure. The count, and average share reward systems used in the dory banks schooner fishery are examined in detail elsewhere (e.g. Andersen 1978, 1980, and 1988).

4   Three kinds of trap fishermen were active on the Labrador coast at this time; permanent settlers ('liveyers'), migratory east coast Newfoundland fishermen working from permanent fishing stations (hence, 'stationers'), or 'rooms,' and migratory east coast schooner ('floater') fishermen. Their season ran from June through about mid-October. The liveyers and stationers usually had customary trap berths near their rooms, while the floaters ranged along the coast with their traps (see Black 1960). Dense caplin schools moving inshore to spawn drew the cod to fill the traps. Trap fishing gave way to cod jigging when the caplin run ended, usually by August, when banker fishermen made their fall trip to the Labrador with squid to bait their trawls. Once on the Labrador, the bankers fished by day from various familiar harbours, and, at night, visited the trap and jigger fishermen and their families ashore (see Hussey's recollections of life at Lear's Room, Labrador, between 1923-47; 1981).

5   I discussed the fate of the *Alsation* with the late Captain George Follett, of Grand Bank, on November 16, 1977. The spring she was lost, he was master of the *Freda M.*, fishing off Sable Island. A storm was breezing up fast one day, and it was snowing, so he was running close-hauled from Sable Island, across the Gully (deep water) to Quero Bank when another vessel passed them under full sail, moving in the other direction. "She was goin' so straight for Sable Island as I am looking at you. I said to Ches' Thornhill, 'Run down and get the flag. In the name of God, get the flag!' (to signal them to come around, change course). It was only minutes, and he come up and said, 'I can't find the flag.' Well, she was gone out of sight, and it was thick-a-snow. Never heard tell of her after." Later, back in Grand Bank, he described the condition of the vessel's sails and color of her dories to the owner, Sam Patten, who said, "There's no mistake about it. That's the *Alsation.*"

6   According to a short note in the St.John's *Evening Telegram* on March 1, 1935 (p. 4), the *Arthur D. Story* cleared Belleoram on February 28, 1935, bound for Gloucester with 700 barrels of salt herring, 100 barrels of pickled herring, and 300 barrels of frozen herring in bulk. She had still not reached her destination or been reported by March 29, when a coast guard alert was announced. The schooner was believed caught in a Gulf of St. Lawrence ice jam (op cit., March 29, 1935, p. 4). Many believe she was overloaded, even her forecastle was filled with herring. "This caused her to set by the head so much that they put her anchors back on the quarters to try to trim her" (Evans 1981, p. 7).

7   The *Laverna* was built and launched in Essex, Massachusetts, in 1911. She was a 141 ton, 106 foot vessel, with a 25 foot beam and 12 foot depth.

After much success fishing, her American owners sold her in January, 1920, to S. Harris, Ltd. of Grand Bank (Thomas 1973, p. 163-164).

8 The St.John's *Evening Telegram* (September 28, 1936, p. 4) carried an item suggesting that an engine spark caused the fire on the *Paloma*. She was beached and her crew landed safely. But her stern was burnt off from back of the afterhatch.

9 It was common practise in October and early November, with the close of the Labrador coastal fishing season, for a banks schooner skipper to provide Labrador seasonal fishermen (chiefly 'stationers') with passage for their men, women and children, trap boats and gear, and catch, to the extent that space permitted. Jam-packed with fish and oil at the close of the season, the floaters could not bring home all the people and gear they had taken to the Labrador in June. The large quantity of barrelled cod oil placed on the *James and Stanley*'s deck, and later jettisoned, was probably purchased from various Labrador fishermen, on consignment to a particular Newfoundland merchant house. The passage under sail from Labrador to Newfoundland east coast ports was often stormy in the fall months. Nicholas Smith (1936, pp. 153-154), for example, recalled a similar passage aboard a banker in October 1934, from Indian Harbour to Cupids, Conception Bay, "...running under bare poles with a man lashed to the wheel," when the puncheons of oil, fishing, and other gear, were swept from the deck.

Chapter 6 — Just a Six Hour Journey

1 In the Newfoundland Supreme Court hearing, Sellars vs. Grand Bank Fisheries Limited, December, 1938, Justice J. Lent determined that the plaintiff was entitled to a "general average contribution" for the goods jettisoned from the deck.

2 The mate with Captain Thornhill on this occasion was Garfield Rogers of Grand Bank. Mr. Rogers recalls there was an eighth man, a passenger, Malcolm Foote, going to Burin. The other crew members were Gordon Hollett, Ambrose Green, Bernard Foote, Wilson (Wils') Green, and the skipper's brother, Wils', who now resides in Sarnia, Ontario.

3 Also "barricade." A small platform about three feet high at the head or bow of the vessel. It enables the crew to manage the jib sail with ease. The term itself may derive from British man-of-war usage. The Oxford Companion to Ships and the Sea (Kemp 1976, p. 62) identifies the barricade as a structure at the forward end of the quarterdeck designed for protection against enemy naval small arms fire.

4 Also called "riding sail."

5 On Friday, December 2, 1938, the St. John's *Evening Telegram* reported:
Yesterday afternoon the Lighthouse Division of the Department of Public Utilities received a message to the effect that the schooner *Allan F. Rose* had been abandoned and the Master, Hubert Vallis, and crew had been rescued by the S.S. Mormae-

sun, bound to Scandinavian ports from New York. The rescue was effected about 250 miles N.E. of Cape Race on November 26th.

The schooner *Allan F. Rose* was built at Belleoram in 1909. She was a vessel of 89 tons and for a number of years has been engaged in the coasting trade. Captain Vallis, who is a well-known master mariner, is the owner of the vessel.

6 The draw bucket usually had about two or three fathoms of line attached to its handle. It had many uses, e.g., to wash down the deck of the vessel and as a toilet. For the latter purpose, it was commonly used at the head of the vessel, for some shelter and a little privacy behind the cable.

7 The St.John's *Evening Telegram* carried the following short news item in the lower right-hand corner of page 4 on Thursday, December 1, 1938:

SCHR. J.E. CONRAD NOW REPORTED SAFE

Fears for the safety of the well-known banking vessel *J.E. Conrad* were allayed when the Captain of the schooner *Calvin Pauline* arrived early this morning at Burin from Sydney with a cargo of coal and reported the vessel, with her sails badly torn, making toward St. Pierre. The *Conrad* was reported shortly afterwards as safe at Grand Bank. She left the latter port on November 24th, for Burin, and got caught in Sunday (Nov. 25) night's gale.

They had met one of the most severe storms to strike the eastern North American seaboard in many years. Wind and snow driven at gale force struck as far south as the Virginia Capes, swept over the New England and eastern Canadian and Newfoundland maritime regions, and left many battered vessels in passing. Around Newfoundland, the *Palfrey*, bound from St. John's to Glovertown, the *Allan F. Rose*, the *J.E. Conrad*, and others, were either forced to run or driven far from their destinations and feared lost when they failed to arrive as scheduled. The Trinity schooner *Marion Rogers* struck the rocks near the lighthouse on Sunday night, November 28, 1938 — no sign of survivors. The schooner *Monica Walters* was lost at Black Island that same night, but the four crew members were safe at Exploits.

Chapter 7 — Never Lost a Line, Never Lost a Hook

1 If one thinks of fish catch alone, what difference did these light-duty engines make? There is no record of when particular banking vessels obtained auxiliary engines, and what catch records that exist for specific vessels are not easily comparable from year to year — depending on how much time was actually devoted to fishing, so it is difficult to say. But we have the detailed log of the *Freda M.*, an eleven dory Grand Bank fishing schooner skippered by the Garnish born highliner, Captain George M. Follett. It records an average yearly voyage of 4,180 quintals for the eight year period 1930 to 1937 inclusive. During the next

three years, 1938 to 1940 inclusive, she averaged 6,185 quintals. If an engine was installed in the *Freda M.* following the 1937 voyage, it may have effected a 50 per cent catch increase. In this case, with a top fish killer, it meant an additional 2,000 quintals annually.
2 "Swiftons Days" is a variant of St. Swithin's Day familiar in the proverb-rhyme, "St. Swithin's day, if thou dost rain, for fifty days it will remain; St. Swithin's day, if thou be fair, for fifty days 'twill rain na mair." Dr. George Story, late author and English scholar at the Memorial University of Newfoundland, provided this interpretation. He suggests that the specific day is 15 July.
3 Andrew Horwood (1971, p. 48) suggests that Captain Alex Smith was known among banks fishermen as "the fish killer." He skippered for G. & A. Buffett, another Grand Bank fishing firm. The *Nina W. Corkum* was under his command for 14 of his 38 years as a banks skipper.
4 According to Allen Evans (1981, p. 9) American-built vessels were low set and deep draft, and drew about 17 feet. "They usually had several tons of pig iron under the flooring of the hold, which made them set lower in the water, making it much easier to hoist out the dories and take them back, also to fork the fish aboard from the dories."
Further, Shelburne-built vessels were like the Americans, while the largest Lunenburg vessels drew but 12 to 13 feet. Compared to those of Lunenburg, the deeper draft vessels were better sailing to windward. Some apparent difference in keel depth was due to bulwark height rather than ballast. By some banks fishermen accounts, many Nova Scotian vessels were simply built with higher bulwarks. One Newfoundland vessel, the *Alberto Wareham,* built for banks dory fishing, had bulwarks so high as to be impractical. She was converted to freighting (cf.H.I. Chapelle, *American Fishing Schooners*).

Chapter 8 — Hang On a Little While
1 The other skipper referred to also worked for Arch's firm.
2 This is probably the same "Olando" Lace, who Captain J. M. Fudge (1963, p. 17) recruited as deckhand ('kedgie') to the crew of the schooner *Harry A. Nickerson* fishing from Gloucester in 1913. Lace may have been 14 at the time.
3 To file a "protest" is to make a formal declaration before an appropriate legal authority, often a local magistrate, about the circumstances of loss or injury to the vessel concerned.

Chapter 9 — In Another Dummy Amongst the Big Power Vessels
1 Senator William Duff (1872-1953), born in Carbonear, Newfoundland, in 1895 settled in Lunenburg County, Nova Scotia, where he became a captain of industry and politics. He was a fish merchant and vessel owner, and became President of the Lunenburg Marine Railway Co., Chester Basin Ship Building Co., and Lunenburg Mutual Marine Insur-

ance Co. From 1904 to 1935 he held various local political offices, and became Liberal Member of Parliament for Lunenburg, and Deputy-Speaker of the House of Parliament. He was called to the Senate in 1936 (Smallwood, Vol. I, 1981:652).

2 Stan, Skipper Arch's younger brother, also recalled the occasion. He said the submarine crew gave their position as 60 miles east of Sable Island. He thought it an English boat. At any rate, Arch sent a dory and two men over. They were given one case each of mutton and cocoa. More interesting is the reason for the encounter: the submariners wanted some halibut! But since Arch's men had not yet begun to fish, the submarine returned that evening for its fish. Now, whose boat was that?

Chapter 10 — A Wonder for Her Time

1 Information attributed to the *Ronald George*'s owners and reported in the St. John's *Evening Telegram* (September 17, 1947, p. 3) suggests she had 120 horsepower when first purchased from Lunenburg in 1942. In 1944 her owners installed a 225 horsepower "Imperial Atlas." Captain Thornhill's estimates of her power remain in this chapter as he recalled them.

2 'Dragging' refers to 'trawling' for groundfish. The boat literally drags its catching gear on the bottom. With the trend toward technological diversification to enable fishing boats to exploit ground and dimersal species, trawlers today often carry both bottom and mid-water trawling gear.

3 Allen Evans (1981, p. 21) tells a humorous story of an old banking skipper and his first radio receiver. A 24-hour marine forecast from a Glace Bay, Nova Scotia, station came on at 1:15 p.m. each day. One day, the skipper, .. turned on his radio in the cabin to pick up the daily forecast, but nothing happened, and although he and the crew members who were gathered around strained their ears, not a sound came forth. He switched it off and on a few times, so since nothing developed he stood menacingly over the set and pronounced in his most commanding tone, "I'm giving you five minutes to speak, boy, and that's final." At the end of the five minutes, he grabbed it off the shelf, went on deck and dumped the thing over the rail just as someone ran up on deck shouting, "Skipper, you didn't have the battery connected to the radio."

4 Campbell Macpherson became Newfoundland's Lieutenant Governor in 1957. Sir Gordon Macdonald was Governor of Newfoundland from 1946 to 1949.

5 On September 17, 1949, the St. John's *Evening Telegram* (p. 3) reported that the *Ronald George* had landed fish worth more than $500,000 during the last five years. Further, "each year both in 'fresh' fishing and 'salt' fishing she has been highliner of the Newfoundland banking fleet,

averaging 6,000 quintals per year. The *Ronald George* has landed over 6,000 quintals this season and is still on the Grand Banks doing well..."

The report also estimated that the individual crew share, exclusive of other bonus, totaled between $1,400 and $1,500 for the season, nearly twelve months of the most dangerous industrial labour.

Chapter 11 — Millions of Fish

1 A distinction is made in Newfoundland between 'baymen' and 'townies.' Baymen or 'bay people' live in the small communities or 'outports' around the province's coastline. Townies live in the major population centres of the island, especially St. John's, the capital city. Like familiar distinction between country and city people, these are significations of space, power, lifestyle, and culture.

2 There is a close and old link between Newfoundland's fishery and emigration of its labour force to mainland North America. David Alexander (1976, pp. 62-63) estimated that export prices for Newfoundland's salt codfish fell some 32 per cent between 1880/84 and 1885/99, while production sagged some 20 per cent and industry gross earnings declined by 36 per cent. Further, the male labour force employed in catching and curing fish fell dramatically, from 60,000 to under 37,000, between 1884 and 1891. The Dominion's net annual immigration from 1884 to 1901 was between 1,500 and 2,500, and from 1,000 to 1,500 in the period 1901 to 1945. Prior to the Second World War, much of this manpower, and that from the Canadian Maritime provinces, immigrated to New England, and especially the 'Boston states.' In the 1890s, back in Newfoundland, many vessel owners reportedly rationalized some of their problems by arguing it was impossible to get "suitable fishermen" for the banks industry, because the best left the country, and those remaining were "not qualified to compete with French and American fishermen" (*Newfoundland Fisheries Commission Annual Report* for 1892, p. 67, 1893). But this was to cite a major symptom of more fundamental economic problems in the saltfish industry.

In the early 1920s, it seems Newfoundlanders and Nova Scotians were in the majority in New England bank fishing crews. "Thus it is that when a Gloucester fishing schooner is lost, mothers and widows in Newfoundland and Nova Scotia remain to mourn the majority of the vessel's dead" (Wallace 1921, p. 5). The situation might have been likewise even in the 1850s (see Innis 1954, p. 333). As late as 1964, almost two-thirds of the crewmen of the Boston-based offshore trawler fleet were born in the "Maritime Provinces of Canada — mainly Newfoundland and Nova Scotia" (Norton and Miller 1966, pp. 7-8). The average age of the Boston trawler labour force was over 55 years then, while it was only about 35 years among Newfoundland's newly expanding offshore trawler fleet.

3 The *Fearless* (later renamed *Blue Foam*), built in 1946, was 140 feet in length, 399 gross tons, and had 805 horsepower. (*International Commission for Northwest Atlantic Fisheries Statistical Report* of 1959, "List of vessels over 50 tons fishing in the ICNAF convention area in 1959.")
4 Perhaps the first attempt to establish groundfish trawlers in Newfoundland was made in 1935. The merchant house of Crosbie and Co., St. John's, incorporated the Newfoundland Trawling Co., Ltd., to operate two trawlers. They were to land fresh cod for curing ashore (*Evening Telegram* February 9, 1935, p. 6). The venture was short-lived, perhaps because of the unexpectedly high threshold costs.
5 This was the 125 foot *Pennyson*, owned and operated since 1947 by George Penney and Sons Ltd. of Ramea (Kendall 1973).
6 This was probably the *Mustang*, a wood hull vessel built at Bay D'Espoir, Newfoundland, and rigged like a sail schooner with two masts. She was about 100 to 110 feet long and had a hold capacity of one hundred to one hundred and twenty thousand pounds. Fishery Products Ltd., her owners in Newfoundland, soon acquired two small — approximately 90 feet long — steel draggers, each with a capacity of about one hundred and fifty thousand pounds.
7 Although not contrary to the pattern of distinct fishing areas outlined by Captain Thornhill for French and English trawlers, and Newfoundland banking schooners, some evidence suggests their relationships were not always harmonious. Allen Evans (1981, pp. 7-8) recounted that the St. Jacques schooner *Marion*, Captain Isaac Skinner, of Boxey, was lost with all 17 hands in March 1915. It was originally thought lost on the St. Mary's Cays. In the early 1930s, however, a Montreal newspaper carried an article that reported that a French beam trawler captain who had had a violent argument (over whether or not the trawler was destroying the bottom) with Captain Skinner in St. Pierre, and threatened to cut him down if he caught him at sea, "had confessed before his death to ramming a banking vessel and massacring her crew."
8 The advent of offshore trawling from St. John's in 1948 gave Newfoundland's government fishery scientists new cause and, via the trawlers themselves, opportunity to gather information about offshore stocks and waters. A result, each trawler carried "trip sheets" on which were recorded date, drag number, position, depth, drag time, catch by species, and other facts pertinent to the fishery. Upon returning to port, a fishery technician made out a "turnout" sheet of actual landings. The original trip and turnout sheets for the *Blue Foam* were located with the help of Nauss Cluett, of Fisheries and Oceans, St. John's. Cluett had sailed on some of the trips recorded.

These data were not located until after Captain Thornhill's death. Save for minor editorial changes and additions from our discussion-interviews and newspaper accounts, this chapter is largely as given me by the Captain shortly before his death in 1976. The trip and turnout sheet records are an opportunity to compare his recollections for consistency. They bring us as close to day-to-day fishing events as we might ever hope to get. Such a comparison underlines the formidable, nay, near impossible task our Captain faced when recalling events and fishing outcomes with precision. He made 38 trips to the banks in 1948 alone, and more than 450 over about fourteen years while dragging. Each trip had an outcome as mixed as the species caught. Data in the Appendix illustrate this difficulty. Figure 1 gives a landings breakdown by species for Captain Thornhill's first three trips in 1948. Figure 2 gives the trip-by-trip landing record for the *Blue Foam* (ex-*Fearless*) and *Blue Spray* (ex-*Challenge*) in 1948. And Figure 3 gives the *Blue Foam*'s annual catch totals and trip averages from 1948 through 1955.

As we might reasonably expect, the Captain's recollections telescope a myriad daily outcomes and trip results for the *Blue Foam*, and other trawlers. The recollections do differ from the trip sheet and turnout record. But comparison also demonstrates his attempt to give an accurate, balanced view of the high and low points of these years. There is little argument with the general accuracy of many points, e.g., the depth and character oft-fished grounds, fish prices that rarely changed over the years, destruction of young fish year after year, sleep lost repairing nets torn on hard grounds, and the radical change in worklife effected by the shift from banks dory schooners to modern trawlers.

9 That is, they had little to show for this fishing effort.
10 A 'set' here refers to the duration the trawl gear is towed. Depending on gear design, bottom and catch, a set may be for as little as about 15 minutes on up to two hours or more.
11 A trip of 300,000 pounds or more was a high point during these early years. They were infrequent, and became more so by the late 1960s, as the international offshore fishing competition intensified in these waters. What did a trip like this in 1948 yield an ordinary deckhand? An actual trip in March 1948 was calculated like this:

Species	Weight Landed	Price/lb.	Total Value
Haddock	217,420	3c/	$ 6,522.60
Cod	76,495	2 1/2c/	1,912.38
Grey Sole	21,330	2c/	426.60
Rosefish	1,285	2c/	25.70
Halibut	4,980	10c/	489.00
Halibut Liver	84	30c/	26.70
Totals:	321,599 lbs.		$9,411.98
(sec)*			

Division of Gross Total:		Food ('grub bill')
Company (63%)	$ 5,929.54	$ 226 = $ 12.50
Crew (37%)	$ 3,532.44	18

Crew: 18 hands
Average Share = $ 193.47 less $ 12.50 (food) = $ 180.97

* = without ice
Note: Prices paid fishermen for each species were "administered" values set by the trawler owner, not actual market prices.

This net settlement of $180 was before taxes, and amounted to about $30 per day for about six 24 hour days at sea — an average trip then — or about $.80 an hour.

12 Newfoundland trawlers sometimes left mark buoys on the banks from trip to trip. They were valuable aids until, aided by the fathometer, the bottom was known in the detail required for effective trawling, and more advanced electronic navigation and fish detection gear was introduced in the early 1950s. The buoyed location might not yield a good catch on each occasion, and, rarely, the buoy fouled the trawler's propeller. But it helped a skipper to minimize gear damage and fish an invisible bottom.

13 The turnout record indicates that the Captain speaks here of the total of what both the *Fearless* and *Challenge* had caught to this time.

14 The cull size for haddock in 1948 may have been 16" (40cm), and the large quantities of small haddock may not have occurred until two or three years later (E. Sandeman, pers. com.). By then, a three-to-one ratio of discarded to kept fish in a given catch was common. At times it was four-to-one or worse. For example, in June 1951, on the *Blue Spray*'s trip sheet No. 19, it was recorded, "In conversation with Captain Vallis (of the *Blue Wave*, another trawler in the Job fleet), I was advised that in one drag he figures that he had short (sic) 30,000 lbs. Haddock, but when they were picked, iced down about 6,000 lbs., and threw away 24,000 lbs. — fish being undersized."

These high discard ratios apply especially to haddock. But cod, plaice, redfish, and flounder were also discarded owing to small size, to be seen floating dead over large areas of the fishing grounds. Still other hundreds of thousands of pounds were lost when trawl nets burst at the surface. The sea always reclaimed what it bore, but it seems there were times when there was so much dead fish rotting on the bottom that it was sometimes noticeable on the surface. The *Blue Foam*'s mate reported during a trip in March 1955, "stench after hauls on St. Pierre Bank. Believed to be caused by so much dead fish on bottom."

15 The *Evening Telegram* sent a reporter, Mark Ronayne, and photographer, Cyril F. Marshall, on a trip aboard the *Challenge* in February. Between March 13 and April 17, 1948, they produced a superbly

illustrated, six part documentary entitled, "Cruise of Trawler *Challenge*." It celebrated both the trawler fishermen's long and dangerous labour, and a "very important new industry."

16 But what if a man were injured and forced to stay ashore for, say, a trip, perhaps for hospitalization? He might still receive a full share, pooled by the entire crew from their next trip. At least this was the reported practise worked out aboard the *Blue Spray* in 1948 (*Evening Telegram* March 20, 1948, p. 8). Such a practise required a stable crew, however, and not all trawlers met this requirement despite limited employment alternatives at the time.

The old "bank fisheries insurance" was under attack in 1947 as "far too small and inadequate." St. John's newspaper editorials argued that "every fisherman should be protected for an amount of at least $2,000 to cover injury or loss of life" (*Daily News*, January 18, 1947, p. 9). Three years later the legislature was still considering such a figure! Chiding his readers' conscience, an editorialist noted,

"...many have never been in a position to carry life insurance, and many widows and orphaned children have had to face the hard knocks of life almost as paupers when the breadwinner has failed to return from the Banks (*Daily News* May 19, 1950, p. 11).

17 Heaven and hell, and worries, are relative. Work on the foredeck of a modern side trawler like the *Blue Foam* imposes extreme risks to life and limb. In *The deep sea fishermen* (1970:25-26), Captain Alan Villiers' description of this work as he saw it aboard British trawlers in the Iceland fishery applies to Newfoundland side trawlers:

The men work with sharp knives. They stand in slippery fishpounds, hour after hour. The foredeck is a fresh fish catching-and-cleaning plant: heavy wires writhe from the great trawl-winch from which the trawl is operated. Giant steel-shod vanes called otter-boards hold the net open on the sea bed: these come aboard with each haul. The trawlerman runs a risk of being minced in the trawl warps..., mashed by the otter boards or pitched overboard while tending them, flung on the slippery, gut-strewn working deck where the sea sweeps always but never enough to wash away all the slime. If he gets in the sea he has little chance of being rescued...

On February 22, 1948, on the 26th tow of his sixth trip, the *Challenge* (later, *Blue Spray*), commanded by Captain Baxter Blackwood, was into a large catch of haddock when near disaster struck them. They had just taken in 1,200 lbs. when, as Captain Blackwood recorded in his trip log, the "Net took charge — two men carried overboard — one of which was injured. Necessitating returning to St. John's immediately. The other rescued without injury."

In *More than just a union* (1985: 206-207), Gordon Inglis' excellent account of the development of the Newfoundland Fishermen, Food and Allied Workers Union (NFFAWU), we learn how in the 1950s and 1960s Newfoundland trawler companies pushed their skippers "to the limit", and they in turn drove their crews relentlessly. A veteran trawlerman speaks:

"...and the captain pushed the men beyond their limit. Keepin' men on deck sixty or seventy hours at a stretch was routine. If you couldn't take that, well, you weren't fishermen. There was nobody to complain to — if you didn't like it you could stay ashore.

There was times you'd hardly get out of your rubber clothes for a week. If you got a few hours to sleep, you'd just kick off your boots and crawl in. One time, I remembers, when the fish was comin' good, we left on a Monday morning and we was back the Friday night. We fished steady from early Tuesday mornin' until Thursday night — sixty hours without a break, only for meals and mug-ups."

Other examples of these realities follow in this chapter.

18 Once again, the Captain speaks of the total catch of both the *Fearless* and *Challenge* to the end of February 1948. The turnout sheets indicate it was about 1,084,900 lbs. and 996,966 lbs., respectively. A total of 2,081,866 lbs.

19 Raw cowhides ('buffalo hides') were spread and fixed to the underside or 'belly' of the trawl's codend — where the fish is gathered by the movement of the trawl — to protect the trawl's mesh from chafing. Some skippers may continue such practice, although others prefer to attach thousands of short strands of poly-filament synthetic fiber to the belly-area for the same purpose.

20 Each summer, great shoals of caplin, a pelagic species, enter Newfoundland's shallow coastal waters to spawn. One of the three major baitfish species (with herring, and squid) in the banks schooner fishery and an important source of revenue to inshore fishermen who caught and sold them to the bankers, caplin are also pursued by cod, the major quarry of Newfoundland's inshore fishery. Shore-based fishermen capture cod with fixed traps, hook and line — long trawls and jiggers, and, more recently, gill nets. Before the collapse of Northern cod in the 1980s, capped by the 1992 moratorium on commercial fishing, summer inshore landings often exceeded the handling capacity of local fish processing plants, especially where a plant operated its own trawler fleet. Faced by a fish glut, plants gave trawler landings priority and turned away inshore landings or sent them to the meal plant.

Either way, this perennial problem reduced the earnings of inshore fishermen. In anger and frustration, they sometimes dumped their catch and underutilized their catch opportunities. The problem persisted into the 1980s even as stocks sharply declined, while, paradoxically, there was (and is) a widely recognized need to "rationalize" excess processing capacity around Newfoundland's coast. Where this excess is determined to exist is an economic, political, and community issue (see C.P. Russell, "A study of the 'Seasonal inshore codtrap fishery'" 1977).

21 The *Blue Foam*'s trip sheets locate the shoal at Lat. 44°30" N., Long. 50°08" W. This seems to be the shallowest location on the Grand Banks (E. Sandeman, pers. com.). But why is it "Bethel" shoal? In the "Journal of Cephas Pearl," the author reports that on August 20, 1882, the Lunenburg schooner *Victor* set its dories on "Jessie Ryder's" shoal. Pearl holds that it was discovered by Ryder of the Halifax schooner *Bethel* in 1845. Pearl gave the shoal a dead reckoning position of Lat. 45°30" N., Long. 45°55" W. Admitting this is about 60 miles further North, our captain's shoal might be named after this Halifax vessel. Alternatively, the shoal might have become a concentration point for many vessels, men busy at the Lord's work. And perhaps vessels anchored there together on a Sunday while their men celebrated the Sabbath with prayer and song. A fisherman's bethel, and a holy place.

22 The Captain's original estimate was six million pounds each during the first six months of 1948. The turnout sheeet record indicates it was closer to four million pounds. The text has been changed accordingly.

23 Indeed, the turnout landing record indicates the *Blue Foam* landed a November trip of 37,626 lbs. in 1948.

24 This seems a rough approximation of what both trawlers landed that year. It is an underestimation if the turnout sheet figures are nearest accuracy. According to the turnout record, the *Blue Foam* landed 5,941,526 lbs. and the *Blue Spray* 5,445,680 lbs. It is still a total of 11,387,206 lbs!

25 The encroachment of trawlers, presumably mainly Canadian at this time, and resultant gear damage, provoked Hermitage Bay area inshore fishermen to press for a twelve mile limit as early as 1947-48. Newfoundland's Minister of Fisheries at the time pointed out, however, that, even if the Canadian government instituted a twelve mile limit for its trawlers, those of other countries could still fish up to the three mile line. A real solution required an international-level agreement. But a fisheries patrol craft was offered as an alternative toward reducing the most aggravating incursions destructive of fishing gear (*Evening Telegram* June 19, 1950, p. 3).

26 This seems a conservative estimate of the total landed by the *Blue Foam* and *Blue Spray*.

27 On June 7, 1950, the *Evening Telegram* (p. 3) carried a short item lauding the *Blue Foam* for landing 355,855 pounds, "the largest catch ever to be

taken from the Banks at one time." Bigger single trip catches have been taken since.

28 Newspaper accounts of this rescue are given in St. John's *Evening Telegram*, November 16, 1950, p. 11; and November 18, 1950, p. 11. In the *Gudrun*'s home port, the Gloucester *Daily News* (November 17, 1950, p. 6) reported:

Charles Poirier, 42 years, of Boston, crew member of the local dragger, *Gudrun*, was severely injured last Tuesday aboard the boat when the latter was fishing on Grand Banks, off Newfoundland. It was reported that Poirier, whose arms were mangled in a winch he was operating, lost one hand, and the fingers off the other hand.

He was transferred Wednesday to the St. John's, N.F., trawler *Blue Foam* and taken by that craft into St. John's for hospitalization, after being treated by medical corpsmen at sea. The medics had sped to the *Blue Foam* aboard a T-boat from the U.S. Naval Dock at St. John's.

29 Despite losing a crew member to serious injury, Captain Johannssen completed the fishing trip with remarkable success. The Korean War was then raging and dominated the headlines in American newspapers, but on November 20, 1950, the Gloucester *Daily News* carried this front-page item: "Hub Beam Trawler Lands Fare of Dabs." The article continued,

Setting what is without doubt, a record in the landing of dabs for one trip, the Boston beam trawler *Gudrun*, Capt. Axel Johannssen, arrived in this port yesterday from two weeks' trip to Grand Banks. Her hail was 315,000 lbs. fresh fish, of which 300,000 lbs. were dabs. Gorton-Pew Fisheries Co. Ltd.... bought the fare, quoting 5 1/2 cents a pound for the dabs.

Never before in the memory of the oldest waterfront habitue, has there been such a tremendous load of flatfish in one fare. The *Gudrun* caught her fare in three days' fishing.

The craft stands to gross stock some $17,000 if the hail stays true at weighout...(and) a possible gross share of some $400 for each member of the crew.

Heretofore, 20,000 lbs. of flatfish was reckoned a large amount to land from a single trip. The catch weight was later adjusted to 230,000 lbs., still perhaps the largest amount ever landed anywhere to that time.

30 But see landings estimate based on 1950 turnout sheets in Appendix. The *Blue Spray*, Captain Baxter Blackwood, reportedly landed eight and one-half million pounds in 39 trips in 1950, and "not a single accident occurred" (*Evening Telegram*, December 16, 1950, p. 3). The estimate has been rounded.

31 Anthropologist Joseba Zulaika presents an excellent account of the

modern Spanish pair seine trawler technology and its fishermen's lives in his *Terranova, the ethos and luck of deep-sea fishermen* (1981).

32 Anxiety about the growing numbers of foreign — especially French, Portuguese and Spanish — fishing boats on the banks was expressed in St. John's' local newspaper editorials as early as 1948. The editorials initially called for "conservation" of the bank fishery and an international accord to control the fishing. But concern for overfishing quickly yielded to control of destructive clashes between new and old catching technologies.

In June 1950, local editorials and news items characterized foreign trawler operations as "Piracy on the Grand Banks," especially for their wanton and callous disregard of banks dory fishermen. The trawlers reportedly destroyed the bankers' fishing gear and uncautiously ploughed the foggy banks heedless of dory fishermen lives. Whereas Newfoundland trawlers reportedly gave the vessels a berth of four to five miles, some foreign trawlers appeared to use the bankers' positions, and navigation lights at night, as likely places to fish, and swept all around them, destroying their gear at the same time (*Evening Telegram* June 12-28, 1950, **passim**).

Local editorials pressed Canada's Parliament to approve the Northwest Atlantic Fisheries Convention signed February 8, 1949, at Washington, D.C. The International Commission for Northwest Atlantic Fisheries (ICNAF) sprang from these pressures. The banks schooner fisherman's problems were not solved by these developments. The industry was dying. At Grand Bank, a local correspondent writing for a St. John's newspaper wrote,

"It's obvious that if we are to remain in this fish business, it must be towards modernization and improvement of the fresh fish industry that we have to shape our course or else...fishing communities like our own will have deserted waterfronts and grass growing in the streets (*Daily News*, May 23, 1950, p. 11).

Overfishing and gear clashes persisted into the 1970s. Canada established a 200 mile management zone in 1977.

33 Some Newfoundland bankers continued to sail until 1955. But men were increasingly reluctant to join and stay with the vessels. They frequently "abandoned" their voyage at mid-season, leaving those who remained to search for replacements. Some suggested that the leavers simply had low income expectations and were satisfied to earn less than a full voyage's income, or that they preferred to rely on government assistance. An *Evening Telegram* (August 21, 1950, p. 4) editorial, however, rightly challenged these views and pointed out that many men left for Lunenburg bankers, where evidence indicated that they could double their earnings.

By February 1952, at least 16 bankers were reported tied up with no prospect of getting crews. Hollett and Sons, at Burin, for example, with five schooners, couldn't crew even one. Never raising questions about the basic structure of the old salt-fish industry, explanations were sought in irregular earnings, the superior attraction of work on draggers, and land-based jobs in filleting plants, on military bases — that offer a chance for unemployment insurance, a support not available to fishermen. And there was the familiar argument that, "Powerful Nova Scotian schooners, offering greater possibilities, have been shipping Newfoundland crews, skimming the cream of the crop of local fishermen" (*Evening Telegram* February 2, 1952, p. 3).

Amid reports of stranded and penniless banks fishermen, the General Secretary of the Newfoundland Federation of Fishermen called for an investigation of living and working conditions on fishing vessels (*Evening Telegram* June 23, 1952, p. 3). It was too late. The future lay with the growing frozen fish trawler industry, not the past, heroic as it was..

34 The St.John's *Evening Telegram* (January 3, 1951, p. 3) carried an account of this event, and reported the cook's death the next day (January 4, p. 5).

35 The *Gudrun* had continued successfully fishing these waters. The Gloucester *Daily News* recorded hails of 278,000 lbs. of mixed fish on 11 December, and 236,000 lbs. on 28 December, 1950.

The Gloucester *Daily News* on January 15, 1951 (pp. 1 and 8) portrayed Captain (Johann) Axel Johannssen as a "highliner over many years," and gave the *Gudrun*'s specifications. She was built at Bath, Maine, in 1928, 114 feet long, 23 feet wide, and 11 feet deep. Small beside the *Blue Foam*'s 140 feet, she was equipped with a 1,200 horsepower GM engine. The U.S. Navy used her as a minesweeper in Iceland waters during World War Two. She was the *Boston College* when Captain Johannssen purchased her after the War. He renamed her after his youngest daughter.

36 Her last message, an S.O.S., was picked up by the New York Coast Guard Division Headquarters on Sunday 15 at 3.24 a.m. "We are sinking. Position Long. 53 degrees, 45 minutes, and Lat. 42 degrees, 38 minutes." They were about 200 miles south of Cape Race, Newfoundland. An intensive air and sea search found no trace of the 114 foot trawler (Gloucester *Daily News* 16 January 1951, p. 1).

She carrried 17 hands when lost. Twelve were married. 35 children lost their fathers.

37 There were two major reports on the *Gundrun*'s loss in St. John's Newspapers. Captain thornhill recalled that one erroneously reported that there was a 40-mile-per-hour northwesterly gale on the banks, when it was actually up to 80 miles per hour (*Evening Telegram* January 15, 1951, p. 3 and January 16, 1951, p. 3).

REFERENCES

Alexander, David — 1976. "Newfoundland's traditional economy and development to 1934". *Acadiensis* (5(2):56-58. 1977. *The decay of trade.* St.John's: Memorial University of Newfoundland, Institute of Social and Economic Research Studies. No. 19.

Andersen, Raoul — 1977. "Bound for Burin". *Newfoundland Quarterly* Vol. 73 (4):17-22. 1978. "The 'count' and the 'share': offshore fishermen and changing incentives," in *Proceedings of the Canadian Ethnology Society meetings*, at Halifax, 1977. Edited by Richard Preston, pp. 27-43. Mercury Series, National Museum of Man, Ottawa. 1980. "Millions of fish", *Newfoundland Quarterly*, Vol. 76 (2):17-24. Reprinted in *Canadian Issues* special on "Canada and the Sea," vol. 3, No. 1 (Spring 1980): 127-139. 1980. "Social organization of Newfoundland banking schooner cod fishery, circa. 1880-1948", in *Seamen in Society*, Proceedings of the International Commission for Maritime History meetings at Bucharest. Paul Adam, editor, pp. 69-81. Paris: np, 1980. 1982. "Recollections of Struggle." *Proceedings of the IVth International Oral History Congress*, at Aix-en Provence, France. Francois Bedarida, et al, editor, pp.175-185. 1988. "Usufruct and contradiction: territorial custom and abuse in Newfoundland's banks schooner and dory fishery". (*MAST*) *Maritime Anthropological Studies* Vol. 1, No. 2: 81-102. 1990. "'Chance' and Contract: Lessons From a Newfoundland Banks Fisherman's Anecdote". In *Merchant credit and labour strategies in the staple economies of North America*. R. Ommer and C. Vickers, editors, pp. 167-182. Fredericton, N.B.: Acadiensis Press.

Binkley, Marian — 1994. *Voices from the Offshore: Narratives of risk and danger in the Nova Scotian deep sea fishery*. St.John's: Institute of Social and Economic Research, Memorial University.

Black, W.A. — 1960. "The Labrador floater-codfishery". *Annals of the American Association of Geography* 50(3):267-293.

Boyd, Cynthia — 1997. ""Come on all the crowd, on the beach!": The working lives of beachwomen in Grand Bank, Newfoundland, 1900-1940." In Candow and Corbin, editors, pp.175-184.

Candow, James E. — 1997. "Recurring visitations of pauperism: Change and continuity in the Newfoundland fishery." In *How deep is the ocean?* James E. Candow and Carol Corbin, editors, pp. 139-160. Sydney, N.S.: University College of Cape Breton Press.

Chapelle, Howard I. — 1973. *The American Fishing Schooners, 1825-1935.* New York: Norton.

Chard, Una — 1974. "The schooner banking industry". Unpublished manuscript. St. John's: Memorial University of Newfoundland.

Dickman, Ilka D., M.D. — 1981. *Appointment to Newfoundland*. Manhattan, Kansas: Sunflower University Press.

Duff, Walter — 1914? "The fisheries of Newfoundland". Lecture delivered in St. John's by Mr. Walter Duff of the Fishery Board for Scotland. St. John's, n.p.

Evans, Allen — 1981. *The splendour of St. Jacques*. St. John's: Harry Cuff Publications.
Fitz-Gerald, Conrad Trelawney Jr. — 1935. *The "Albatross": The biography of Conrad Fitz-Gerald, 1847-1933*. Bristol: Arrowsmith.
Fizzard, Garfield — 1988. *Master of his craft: Captain Frank Thornhill*. Grand Bank, Newfoundland: Grand Bank Heritage Society. 1987. *Unto the sea: A history of Grand Bank*. Grand Bank, Newfoundland: Grand Bank Heritage Society.
Fraser, A.M. — 1946. "Fishery negotiations with the United States". In *Newfoundland: Economic, diplomatic and strategic studies*. R.A. MacKay, editor, pp. 333-410. Toronto: Oxford University Press.
Horwood, Andrew — 1971. *Newfoundland ships and men*. St. John's: The Marine Researchers.
Hussey, Greta — 1981. *Our life on Lear's Room, Labrador*. Port de Grave, NF: G. Hussey.
Inglis, Gordon — 1985. *More than just a union: The Story of the NFFAWU*. St. John's: Jesperson Press
Innis, Harold A. — 1954. *The cod fisheries: The history of an international economy*. Revised ed. Toronto: University of Toronto Press.
Kelland, Otto — 1984. *Dories and dorymen*. St. John's: Robinson-Blackmore.
Kemp, Peter, ed. — 1976. *The Oxford companion to ships and the sea*. London: Oxford University Press.
Kendall, Victor — 1973. "The fisheries of Ramea". Unpublished manuscript. St. John's: Memorial University of Newfoundland.
Leather, John — 1970. *Gaff rig*. London: Adlard Coles, Ltd.
Macdonald, David — 1980. *'Power begins at the cod end.'* St.John's: Institute of Social and Economic Research, Memorial University.
May, W.E. — 1973. *A history of Marine navigation*. Henley-on-Thames: G.T. Foulis.
Macdermott, Rev. H.J.A. — 1938. *MacDermott of Fortune Bay, told by himself*. London: Hodder & Stoughton Ltd.
Norton, V.J. and M.M. Miller — 1966. An economic study of the Boston large-trawler labour force. *Bureau of Commercial Fisheries*. Circular 248. Washington, D.C.: Fish and Wildlife Service, United States Department of the Interior.
Peacock, Kenneth (compiler and editor) — 1965. *Songs of the Newfoundland outports*. 3 vols. Ottawa: National Museum of Canada. Bulletin No.197, Anth. Series No.65.
Pearl, Cephas — c. 1883. "Journal of Cephas Pearl." Unpublished. Public Archives of Nova Scotia, MG7, No. 14A.
Pennanen, Gary — 1979. "The Fortune Bay Affair, 1878-1881: Massachusetts fishermen versus the British Crown". *American Neptune* 39 (4): 289-301.
Pullen, H.F. — 1967. *Atlantic schooners*. Fredericton, N.B.: Brunswick Press.
Ronayne, Mark — 1948. "The banks trawler fishery provides the nucleus for a very important industry". *Evening Telegram* 17 April: 8-9.
Rowe, Frederick W. — 1980. *A history of Newfoundland and Labrador*. Toronto: McGraw-Hill Ryerson.
Russell, C.P. — 1977. "A study of the seasonal cod trap fishery glut". n.p.

Sager, Eric — 1981. "Newfoundland's historical revival and the legacy of David Alexander". *Acadiensis* XI (1): 104-115.

Sider, Gerald M. — 1986. *Culture and Class in Anthropology and History: A Newfoundland Illustration*. Cambridge University Press.

Smallwood, J.R., ed. — 1981. *Encyclopedia of Newfoundland and Labrador*. Vol. 1. St. John's: Newfoundland Book Publishers (1967) Ltd.

Smith, Nicholas — 1936. *Fifty-two years at the Labrador fishery*. London: Arthur H. Stockwell, Ltd.

Story, G.M., W.J. Kirwin, & J.D.A. Widdowson, eds. — 1990. *Dictionary of Newfoundland English*. Second ed. St. John's, Newfoundland: Breakwater Books.

Sweeney, Robert C.H. — "Accounting for Change: Understanding Merchant Credit Strategies in outport Newfoundland." In Candon and Corbin, editors, pp. 121-138.

Thomas, Gordon W. — 1973. *Fast and able: Life stories of great Gloucester fishing vessels*. Gloucester, Mass.: Gloucester 350th Anniversary Celebration, Inc.

Tunstall, Jeremy — 1962. *The fishermen*. London: MacGibbon and Kee.

Villiers, Alan — 1970. *The deep sea fishermen*. London: Hodder and Stoughton.

Wallace, Frederick William — 1921. "Life on the Grand Banks: an account of the sailor-fishermen who harvest the shoal waters of North America's eastern coasts". *National Geographic* Vol. 40 (1): 1-28.

Wareham, Wilfred W. (ed.) — 1975. *The Little Nord Easter: Reminiscences of a Placentia bayman*. By Victor Butler. St. John's: Memorial University of Newfoundland Folklore and Language Archive. Community Studies No. 1.

White, Marilyn — 1974. "The youth of a schooner banking fisherman". Unpublished manuscript. St. John's: Memorial University of Newfoundland.

Zulaika, Joseba — 1981. *Terranova: The ethos and luck of deepsea fishermen*. St. John's: Memorial University of Newfoundland, Institute of Social and Economic Research. Studies No. 25.

APPENDIX

Blue Foam catch performance.

Fig. 1. Landings by species, first three trips, 1948.
Fig. 2. Comparison of **Blue Foam** and **Blue Spray**, landings by trip, 1948.
Fig. 3. **Blue Foam** annual catch totals and trip averages, 1948-1955.

Fig. 1. **Blue Foam** landings by species, first three trips (Jan. 1948).

Species	Trip 1	Trip 2	Trip 3
haddock	38,284*	126,180	100,575
cod	7,865	7,580	9,655
grey sole	3,005	600	3,050
rosefish	1,540	610	40
halibut	65	481	165
hake	2,945	—	—
'trash' or 'scrap'	8,780	26,000	60,000
TOTAL	62,484	161,451	173,485

* In hundred pounds.

Source: Turnout sheets, Newfoundland Fisheries Research Station, Fisheries Research Board of Canada, St. John's.

Fig. 2. Comparison of **Blue Foam** and **Blue Spray**, trip landings, 1948.

	Blue Foam	(ex-Fearless)	Blue Spray	(ex-Challenge)
Trip No.	Duration	Landings*	Duration	Landings*
1	?	62,484	?	117,470
2	?	161,451	?	118,820
3	?	173,485	22-01:28-01	141,110
4	01-02:08-02	211,790	30-01:06-02	112,842
5	10-02:19-02	72,385	07-02:15-02	133,750
6	20-02:25-02	205,930	16-02:22-02	140,195
7	26-02:06-03	197,375	24-02:01-03	232,735
8	11-03:18-03	287,035	06-03:14-03	238,390
9	20-03:24-03	131,505	17-03:23-03	226,020
10	28-03:04-04	321,599	28-03:02-04	171,000
11	07-04:12-04	195,810	04-04:09-04	185,505
12	13-04:19-04	217,395	10-04:15-04	180,420
13	20-04:26-04	255,598	17-04:21-04	236,285
14	29-04:03-05	142,550	23-04:29-04	144,485
15	08-05:11-05	177,880	05-05:10-05	198,620
16	14-05:20-05	193,820	12-05:17-05	127,295
17	21-05:28-05	186,605	18-05:24-05	226,565
18	30-05:03-06	171,580	27-05:30-05	201,435
19	06-06:12-06	134,945	? :08-06	49,535
20	18-06:20-06	173,830	? :18-06	109,525
21	25-06:28-06	112,765	18-06:24-06	177,850
22	01-07:05-07	211,141	27-06:02-07	98,795
23	11-07:14-07	133,344	18-07:23-07	232,828
24	04-08:08-08	157,255	30-07:03-08	130,380
25	10-08:16-08	168,050	06-08:12-08	110,635
26	18-08:22-08	211,980	13-08:18-08	170,640
27	25-08:30-08	170,065	20-08:25-08	202,095
28	01-09:09-09	161,815	27-08:01-09	126,255
29	10-09:16-09	121,285	05-09:12-09	215,365
30	17-09:23-09	166,945	16-09:23-09	178,560
31	25-09:01-10	51,605	23-09:01-10	90,360
32	02-10:10-10	119,035	23-11:09-12	189,955
33	12-10:19-10	119,655	09-12:17-12	155,200
34	19-10:29-10	159,785	28-12:04-01	219,135
35	31-10:09-11	68,299		
36	17-11:21-11	37,625		
37	23-11:29-11	101,980		
38	03-12:09-12	149,075		
TOTALS (lbs.)		5,941,526 lbs.		5,445,680

* Includes 'trash' fish

Source: Turnout sheets, Newfoundland Fisheries Research Station, Fisheries Research Board of Canada, St. John's.

Fig. 3. **Blue Foam** annual catch totals and trip averages,1948-1955

Year	No.trips	Total Catch (in lbs.)	Avg./trip
1948	38	5,941,526	156,356
1949	30	6,028,401	200,947
1950	37	7,639,268	206,467
1951	37	8,102,145	218,977
1952	37	7,267,300	196,414
1953	29	5,919,993	204,138
1954	30	6,977,327	232,578
1955	32	6,456,527	201,766

Source: Turnout sheets, Newfoundland Fisheries Research Station, Fisheries Research Board of Canada, St. John's.

INDEX

Vessels

Admiral Dewey, 180
Albatross, 297n
Albert J. Lutz, 303n
Alberto Wareham, 315n
Allan F.Rose, 52, 163, 313n, 314n
Alma Harris, 57
Alsation, 130, 145, 180, 312n
Arthur D.Storey, 34, 145, 312n
Baccalieu, 195
Bella Bina P. Domingos, 61, 88
Bessie MacDonald, 25, 63, 65, 69-83, 96, 179, 301n
Bethel, 323n
Blue Foam (ex-Fearless 260), xv, xvi, 4, 8, 203, 241-79 *passim*, 282, 284-86, 289-91, 293, 318-319n, 321n, 323-324n, 330-333
Blue Mist II, 292
Bluenose, 54
Blue Spray (ex-Challenge 260), 260-75 *passim*, 275, 319-24n, Appendix, 330-31
Blue Wave, 291-92, 320n
Calvin Pauline, 314n
Carrie & Nellie, 85, 90-91
Challenge, 248-51, 256, 259-60, 320-22n
Christie & Eleanor, 308n
Clara B. Creaser, 121, 309n
Clarenville, 231, 238
Columbine, 23
Corenzia, 121
Coral Spray, 180
Eileen C. Macdonald, 203
Eva A.Culp, 172
Dauntless, 69, 88-90, 180
Dorothy Melita, 92-101, 102, 141
Eleazor Boynton, 306n
Elsie, 44, 299n
Fearless, 241, 247-260, 317n, 320-21n
Flora S.Nickerson, 6, 89, 304n
Florence xiv, 167, 169-90, 215
Frances, 54

Freda M., 309n, 314n
Freestead, 101-02
Flying Cloud, 247
Garnish Queen, 180
General Byng, 311n
General Currie, 311n
General Horne, 88, 89
General Trenchard, 311n
Gladiola, 180
Glencoe, 88, 93, 107, 308n
Governor Anderson, 230
Grand Prince, 2, 4
Gudrun, xix, 269-72, 276-77, 324n, 326n
Harry A.Nickerson, 315n
Henry Stone, 209-10
Hurricane, 303n
Investigator II, 251, 261
Isabel Corkum, 254
James & Stanley, 88, 130, 139-49, 151, 154, 222, 270, 313n
J.E.Conrad, xiv, 149-69, 290, 314n
L.A.Dunton, 232, 311n
Laddie, 304n
Laverna, 141-43, 147, 312n
Leopard, 180
Lillian M. Richards, 311n
Loch Lomond, 56
Louie H., 21
Maggie Dunford, 301n
Mahaska, 187
Makkovik, 4, 175, 292-93
Mallakoff, 182
Mamie & Mona, 23, 25-6, 45, 52, 297n
Marconi, 24, 54
Marion, 318n
Marion Rogers, 314n
Marshall Frank, 203
Mary Ann, 305n
Mona & Minnie, 297n
Mona & Memmie, 297n
Monica Walters, 314n
S.S.Mormaesun, 313n
Mustang, 250, 318n

N.Fabricius, 73
Nina W.Corkum, 171, 315n
Ocean Ranger (oil rig) 291
Pauline C. Winters, 311n
Palfrey 314n
Paloma, 120-21, 147, 308n, 312n
Pan American, 191-210, 215, 284
Partanna, 130, 147
Pennyson, 318n
S.S.Portia, 102
Preceptor, 88
S.S.Restigouche, 299n
Rex, 97, 305n
R.L.Borden, xiv, 128, 130-40, 208
R.L.Conrad 98
Robert Esdale, 230, 235
Robert Max, 123-26, (ex-Clara B.Creaser) 309n
Ronald George, 212, 214-41 *passim*, 244-46, 252, 257, 264, 287, 291, 316n
Santa Elisa, 273
Santa Maria, 273
St.Ritchards, 275-76
Sanuand, 42, 298n
Stanley & Frank, 25-6, 56, 57
Thistle, 54, 300n
Triumph, 303n
U-Boat #156, 303n
Vera P.Thornhill, 102, 106-18, 120, 128, 150, 193, 195
Victor, 323n
Wally G., 81
Warren M. Culp, 58, 87
Wymota, 213-214
Zerda, 266
Zibet, 275

Industry Individuals and Firms

Adams, Capt., 101
Alexander, Capt., 163
Anstey, Capt., 88
Ayre's & Sons, Ltd., 299n
Barnes, Capt., 54
Barnes, John Capt., 300n
Barnes, John, 236
Bartlett, Sam Capt., 304n
Bartlett, Tom Capt., 310n
Blackwood, Baxter Capt., 245-248-51, 258, 260, 264-5, 321n
Blandford, Capt., 308n
Bonavista Cold Storage Co., Ltd., 3, 292
Bond, Thomas Capt. & merchant, 29, 35, 52, 55-6
Broydell, Arch Capt., 258, 275
Buffett, G. & A. merchant, 309n, 315n
Bullen, Joe Capt., 300n
Burke, John Doctor, 152, 215
Carr, Percival merchant mgr., 120, 131, 154
Cheeseman, V.T. merchant, 301n
Cleveland, Edward Capt., 203, 206
Cluett, Naus, 318n
Coaker, William F., 304-5n
Crosbie & Co., Ltd., 217, 234-6, 318n
Dobbin, Dick Capt., 247
Doyle, Gerald S. News Bulletin, 215, 234
Duff, William Senator, 193, 199, 315n
Elms, Reuben Capt., 58
Evans, Bill Capt., 90
Fishery Products Ltd., 266, 318n
Fitz-Gerald, Conrad Doctor, 296-7, 304n
Fizzard, Mr. George, 64-5, 67-9, 71
Follett, George Capt., 254, 309n, 312n, 314n
Foote, Bernard, 313n
Foote, Malcolm, 313n
Forsey, Curt merchant, 311n
Forsey, William merchant, Grand Bank, 90, 302n
Forward and Tibbo, merchants, 98, 125, 209
Fudge, J.M. Capt., 297n, 302n, 315n
Gorton-Pew Fisheries Co. Ltd., 324n
Grand Bank Fisheries, Ltd., 128
Grandy & Sons, Ltd., 231, 238-9
Green, Ambrose, 295n, 313n
Green, Wilson Capt., 45, 52, 85, 92
Green, Wilson, 313n
Hackett, Capt., 34

Harris, Eli merchant, 301n
Harris, Jim Capt., 301n
Harris, Samuel J. merchant, 61, 69, 71-2, 89, 301n, 309n, 311n, 312n
Harris, Sidney Capt., 203
Harvey & Company, Ltd., 169
Hatch, Freeman Capt., 2
Herridge, Myril, 284
Hollett, Gordon, 313n
Hollett, Murley Capt., 164
Hollett & Sons Ltd., 326n
Hyde, Joy Capt., 99
Job Bros. & Co. Ltd., 220-1, 223-4, 238, 241, 245, 252, 256
Johannsen, (Johann) Alex Capt., 272, 277, 324n, 326n
Johnstone, Phillip Capt., 57
Lace, Orlando Capt., 187, 315n
LaFosse, Sam, 135
Lee, Harry Capt., 65
Levi, Angus Capt., 311n
MacDermott, Hugh Rev., 293, 296n, 299n, 300n, 301n, 304n
MacDonald, Gordon Governor, 316n
MacPherson, Campbell Lt. Governor, 232, 316n
Matthews, Joshua Capt., 121
Matthews, Morgan (Mawg') Capt., 228-30
Miles, Bill Capt., 306n
Miles, Mr., 96
Moulton, Bobbie Capt., 258, 260
Moulton, Wilbert Capt., 81
Myles, Ab' Capt., 203
Newman, Hunt and Co., 297
Osborne, Wilson blacksmith, 3, 12, 303n, 305n
Parrott Brothers merchants, 54;
Parsons, Charlie, 236
Patten, Charlie sailmaker, 3
Patten, Howard, merchant, 162
Patten, John B. (John Ben) & Sons, merchant, 92, 98-9, 101, 145, 149, 154, 162
Patten, Samuel, 312n
Penney & Sons, Ltd., 318n
Pettitte, Gerry merchant, 306n
Poirier, Charles, 324n
Pope, Albert Capt., 57, 91-2

Pope, Bill mate., 284
Pope, Harry Capt., 89
Pope, Leslie, 299n
Price, John Capt., 25-6
Riggs, William (Billy), 3, 175-6, 185
Rogers, Capt., 89
Rogers, Garfield mate, 295n, 301n, 313n
Rose, Charlie Capt., 118, 131
Russell, Roy manager, 257
Sandeman, E., 320n, 323n
Scott, Arthur Capt., 69
Skinner, Isaac Capt., 318n
Smith, Alex Capt., 85-7, 171, 232, 234, 304n, 315n
Snook, Ben Capt., 209
Steers Bros. Ltd., 209
Sullivan, Karl, 6
Thomasen, Harry Capt., 309n, 310n
Thornhill, Arch Capt. (1901-1976): acquires accordion, 60; at St.John's, in 1975, 6; at Grand Bank for first voyage (1918), 69-83; authority and patron, 4-5; born at Anderson's Cove, 18; buys Grand Bank house, 120; captaincy requirements, 103; Christmas, 22, 32; collaboration with author, 7; cousins are captains, 45; contributes lumper wages to parents, 41; death interrupts present volume, 8, 281-2; denied permission to ship out, 51-3, 60; on dying, 282: eager to go deep sea fishing, 28, 39, 42, 45-6; eager to become skipper, 91-2; education, 19; elder brother joins British navy, 62; elder brother ships out, 45; elder brother helps support family, 50; elder brother returns from war, 85; family origins, 17-8; father an eager inshore fisherman, 50, 59; father goes deep sea fishing, 50; first command, 106-19; first dory berth, 64-5; first trip on sailing vessel, 42; first time as schooner mate ('second hand'), 93-4; fishes inshore with family, 43, 53; friendship at Stone's Cove, 22, 91-2; game warden assistant, 43; his large family,

59; hunting as necessity, 53; illness (blood poisoning) in youth, 20; interviews and issues, 7; lumper, 41-2, 45, 53; learns banks trawl fishing from elder brother, 59; meets new teacher, 83; moves to St.John's, 258-9; obligation to employer, 99; patience, 7; purpose in telling his story 17; registers for school, 21; reputation, 287-8 resists seasickness, 40, 42, 73-8, 82, 84; second phase of career in command, 140-; trawler command, 245-79; youth at Anderson's Cove, 18-9

Thornhill, Catherine, 245, 279, 281, 290

Thornhill, Charles (Charlie), 9, 25, 31, 43, 44-5, 49, 52, 61, 299n

Thornhill, Chesley, 312n

Thornhill, Cyril (Arch's son), 251, 260, 284, 289-90

Thornhill, Cyril (Arch's brother), 281, 299n,

Thornhill, Eli James, 45, 50-2, 61-2, 85

Thornhill, Florence, 245, 259, 288, 290, 293

Thornhill, Frank Capt., 94-5, 141-3, 147, 172, 296n

Thornhill, Jacob Capt., 230

Thornhill, John T. Capt., 101, 103, 106, 120-1, 123-5, 305-6n, 308n, 309n

Thornhill, Reuben (Reub') Capt., 92-9, 237

Thornhill, Roland, 259-60, 281, 290

Thornhill (nee Williams), Ruth (Mrs.): career, 45; hospitality 6; responsibilities, 104-5; supportive role 7-8, 106, 118-9, 133-4, 162-3, 259-60, 275, 285, 290, 306n, 308n

Thornhill, Stanley, 61, 283, 298n, 305n, 316n

Thornhill, Wilson (Wils'), 123, 163, 313n

Thornhill, William Capt., 172

Tibbo, James Capt., 23

Tibbo, Charlie merchant, 209

Vallis, Hubert Capt., 313-4n, 320n

Wareham, Frederick merchant, 230, 238, 246

Wareham, Harry merchant, 223

Wareham, Henry merchant, 238

Wareham, Leman merchant, 238

Wareham, W.W. merchant, 211-2, 214-5, 223-6, 230-1, 238, 245

West, Waldram, mate, 217

White, Chesley (Ches'), 87

Williams, Arch telegrapher, 297n

Williams, Clarence (Clar) Capt., 133-5, 282, 304n, 311n

Williams, Gordon Capt., 133-5

Williams, Thomas H. merchant, 56

Other individuals

Alexander, David, 296n, 305n
Bernard, Steven, 299n
Binkley, Marian, 296n
Black, W.A., 312n
Boyd, Cynthia, 295n
Candow, James, 295n
Chapelle, H.I., 315n
Conrad, Joseph, 153
Corbin, Carol, 295n
Dickman, Ilka, 308n
Duff, Walter, 306n
Duncan, Norman, 283

Evans, Allen, 297n, 304n, 306n, 309n, 312n, 315n, 316n, 318n
Fizzard, Garfield, 296n, 301n
Fraser, A.M., 298n
Horwood, Andrew, 303n, 315n
Horwood, Harold, 309n, 310-11n
Hussey, Greta, 312n
Inglis, Gordon, 296n, 321-2n
Kelland, Otto, 304n
Kemp, Peter, 313n
Kendall, Victor, 318n
Leather, John, 304n
Macdonald, David, 296n

Marshall, Cyril F., 320n
Miller, M.M., 317n
Murray, Heather C., 295n
Norton, V.J., 317n
Peacock, Kenneth, 305n
Pearl, Cephas, 323n
Pennanen, Gary, 297n
Pullen, H.F., 298n
Ronayne, Mark, 320-1n
Russell, C.P., 323n

Sager, Eric author, 283
Sider, Gerald, 308n
Smith, Nicholas, 304n, 312n
Story, George, 315n
Sweeney, Robert C.H., 308n
Thomas, Gordon W., 305n
Villiers, Alan Capt., 321n
Wallace, William, 303n, 307n
Wareham, Wilfred, 308n, 317n
Zulaika, Joseba, 324-5n

Places mentioned

Anderson's Cove, Fortune Bay, 15-65; abandoned in 1960s, 20; church services 19; American schooners bring produce, trade gold for herring, leave scaffold wood 34; berry picking 32-3; boys work as lumpers 41; credit relations with merchants; fishing location for salmon 15; baitfish (herring) location 17; bottled foods 27; conflicts with American fishermen over herring stocks 15; caplin runs 28; cash for church donation 29; children's work 30; cleared out when its men depart for banks fishing 61; coal for home heating 29; coastal traders 31-2; cod liver oil 30-1; conflict (youth) with neighbouring outport 21-2; construction of Congregational church 20; credit relations with merchants 30-1, 33, 35; dole 27; dogs and sleighs 29; education 19, 43; first Congregational teacher 20; first local men go deep sea fishing 16, 27-8; on shore 'fishmaking' (cod), 30, 40-1; full use of fish 30; game and other resources 16, 18; gardening 16; frontier resources under pressure 35; gender roles 30-1, 46-7; grass for hay 28; homecoming celebration 24-5; happy home 27, 32; herring trade 33-4; house design 27; household production and self-sufficiency 16; hunting 28, 53, 59; illnesses: diptheria 62-4; inshore fishing 27, 32, 46-7; inadequacy of fish production alone 31; integrative community occasions 16; intermarriage between outports 22; kinsmen on area schooners 48; knitwear 28; lifestyle and work regime 15-6, 28; livestock 28; lobster fishing 28, 46-7; lobster processing and canning 47; local trader 44, 49; luck 29; mothers make family clothing 59; population 15, 26; quarantine 64; relations with Stone's Cove people 21-2; mourning and loneliness 24, 26-7; New Year's festivities 59, 62; produce from Prince Edward Island 28; 'splitting fish' 30-1; religious affiliation 19; root cellar 27; salmon fishing 44; settlement 16, 18, 20; safe harbour 17; selling caribou meat for cash 29; seasonal fishing regime 30; starvation fear 49, 53, 92; tranquility 18; winter houses 29; women form berry picking crews 32-3; women trade own production (cod liver oil and berries) 31-3; woodcutting 33; World War I interrupts lobster trade 49; values 16, 27; vegetable storage 27; visiting sportsmen 44; other, 107, 121, 150, 163

Arctic ice, 92
Argentia, 200, 269-72
Aquaforte, 79
Baine Harbour, Placentia Bay, 147, 220

Barbados, 118, 129-40, 208, 275, 310n
Bath, Maine, 326n
Battle Harbour, 55
Bay D'Espoir, 126, 248, 299n, 318n
Bay L'Argent, 31, 54, 217
Bay of Islands, 34
Belleoram, 19, 23, 28, 29, 35, 45, 52-4, 88, 129, 169, 297n, 300-1n, 302n, 312n, 313n
Belle Isle Strait, 148
Berea, Kentucky, 106
Bermuda, 5
Bonne Bay, 54
Boston, 244-5, 247-8, 324n
'Boston States' (i.e. New England), 129-30
Boxey, 318n
Brazil, 124
Bridgewater, Nova Scotia, 111
Brigus, 304n
Britain, 193
Brunet Island, 25
Burgeo, 93, 131, 139
Burin, 10, 107, 112-3, 141, 153-65, 248, 250, 254, 266, 314n, 326n
Campbellton, New Brunswick, 129, 134-6
Calvert, 79
Canada, 37, 44, 129
Canso, 135
Cape Anguille, 54
Cape Broyle, 79
Cape Chapereau, 157
Cape George, 54
Cape Harrison, 143
Cape Race, 46, 131, 147, 151
Cape Ray, 46, 53
Cape St. Mary's, 230
Carbonear, 193, 303n, 315n
Clattice Harbour, Placentia Bay, 220
Codroy, 54
Conne, 44
Cow Head, 54
Cupids, 313n
Dantzig Point, 154
English Harbour East, 54, 299n, 306n
English Harbour West, 17, 30
Epworth, 90-1
Essex, Mass., 312n

Femme, 29, 44, 60
Ferryland, 79
Fortune, 55, 129, 131
France, 33, 192
Funk Islands, 148, 154
Garnish, 87
Gaultois, 53
Gloucester, 33-4, 50, 113, 145, 199, 269-72, 312n, 315n, 324n, 326n
Grand Bank, 3-257 *passim*, busy in spring, 69; dancing prohibited, 79; busy after spring trip, 78-9; ice delays arrival of returning foreign-going vessels, 92; local men lost when bowsprit submerged, 88; order despite many fishermen in port, 79; shipyard, 78; shore fishers, 79
Green Island, 154, 157, 164
Greenland, 193
Gulf of St.Lawrence, 92,
Gulf Stream, 137
Harbour Breton, 30, 35, 147, 297n, 304n
Harbour Grace, 217, 234-5, 237-8
Harbour Le Cou, 57; vessels jammed in Gulf ice, 90
Halifax, 23, 92, 110-11, 113, 117, 124, 129, 131, 188, 195
Harbour Buffett, 214, 217-21, 223-4, 227, 230-1, 235, 237-8, 241
Harbour Mille, 29, 54
Holyrood, 112
Hoop Cove, 20
Indian Harbour, 313n
Indian Tickle, 55
Isle aux Morts, 213
Jacque's Fontaine, 57; dory crew lost, 85
Jersey Harbour, 118,
Labrador Coast, 43, 303n
La Have, 106-7, 111, 113, 117-8, 149-50, 194
Lamaline, 113, 158
Lawn, 97
Lear's Room, Labrador, 312n
Little Bay East, 19, 29, 42, 45, 49, 52, 54, 56, 69, 85, 96, 106, 306n
Lost Child breaker, 156

Louisbourg, Nova Scotia, 110
Lunenburg, Nova Scotia, 33-277 *passim*
Marstal, Denmark, 73
Marystown, 10
Montreal, 209, 293
Mose Ambrose, 309n
Mount Carmel, St. Mary's Bay, 247
New England (also 'Boston States'), 179
North Harbour, Placentia Bay, 220
North Sydney, 22, 29, 50, 80, 101, 106, 110, 118, 121, 129, 187, 224, 237
Nova Scotia, 26, 106-7, 112, 119, 226
Oporto, 92
Pool's Cove, 56, 62, 83-5, 106, 297n, 299n, 308n, 311n
Port au Bras, 301n
Port aux Basques, 102, 220, 256-8, 260
Portugal, 61; market slump, 89; vessel lost returning from, 89
Prince Edward Island, 23, 28-9, 43, 50, 97-9, 124, 127-9, 131, 220, 224, 235
Pushthrough, 92, 121
Ramea, 318n
Recontre East, 24, 54, 88, 106-7, 150, 187
Rencontre West, 92, 308n
Riverport, Nova Scotia, 121
Rose Blanche, 45, 92
Sable Island, 136

Sackville, New Brunswick, 247
Sagona, 89
Salmon Bight Passage, 147, 181
St. Anthony, 54
St. Bernard's, 54
St. Jacques, 20, 30, 54, 62, 297n
St. John's, 6-256 *passim*
St. Mary's Bay, 54, 81
St. Mary's Cays (Keys), 161, 318n
St. Pierre and Miquelon, 54-318n *passim*
Southern Shore, 79-80; ice houses, 80; shore fishers sell caplin and ice to schooners, 80; 287, 303n
Sarnia, Ontario, 313n
Sheet Harbour, Nova Scotia, 209
Shelburne, Nova Scotia: schooner source, 102
Stone's Cove, 19-101 *passim*
Svenborg and Omegns Museum, Denmark, 73
Sydney, N.S., 124, 164, 188, 235, 262, 314n
Terrenceville, 53
Trepassey, 24, 55, 131
Turk's Island, 138, 275
Western Marine Railway dockyard, Burin, 155
West Indies, 140
U.S.A., 23, 24, 37, 44, 50, 248
Woody Island, 34

General Subjects

Accidents: collisions, 56-7; dory, 8-9, 43, 26, 72; groundings, 21, 23, 87, 113; on board, 56; storms, 21, 23, 57; signal cannon explosion, 11-11; August gale, 97
American vessels, 54, 78, 167, 169, 297-8n
Amulree Report, 309n
Atlas Imperial engines, 43, 224, 316n
Avalon School Board, 299n
Bait:herring, 15, 16, 17, 25, 33-4, 46, 50, 54, 300n, 312n; caplin, 28, 49, 54, 79-80, 217, 303n, 322n; mackerel, 217, 237; squid, 30, 80, 92-3, 112-3, 182, 184
Barter trade, 33
Bays: Conception Bay, 148; Connaigre Bay, 72; Fortune Bay, 4-223 *passim*; Hermitage Bay, 54, 81; Placentia Bay, 10, 54, 79, 81, 92, 112, 161, 220, 235; St.George's Bay, 120; St.Mary's Bay, 54, 81; Trinity Bay, 182
Belonging, 37
Berry-picking, 32-3
Boarding house, 3, 106, 299n
Burin Peninsula, 3, 5, 153

339

Canadian National Railway, 213, 224, 231, 261
Caribou, 28-29
Cash scarcity, 29-34, 89
Change: technology and work values 2; society, 241; work 9
Chester Basin Ship Building Co., Nova Scotia, 315n
Children: berry picking, 32-3; learning skills, 16; work beside parents, 16; hunting (boys), 28; (cod) 'fishmaking', 30, 42; help in lobster fishery (boys), 46; roles, 36-7, 40-1; when boys seek "work", 41; woodcutting (boys), 32
Christmas, 22-268 *passim*
Class differences, 28-9, 36, 167-8

Co-adventurers, 10 (see Share systems)
Coastal passage, 24-235 *passim*
Coastal traders, 31-3, 43, 44, 299n
Coast Guard, 269-72, 326n
Communication:58;radio, 162-3, 203, 212, 235 (G.S.Doyle News Bull.); radio-telephone, 227-8;telegraph, 20, 24, 25, 52, 61-2, 77, 88, 101, 162, 189, 197, 199, 245
Community values, 167-8
Confederation, 213, 241
Credit relations with merchants: 30-1, 44, 49, 68, 116-7, 120-1, 218, 299n, 307-8n, 309-10n
Customs (Service), 124, 139

Deep sea fishing I:
Banks dory schooners (67-239)

alternative employment, 105, 126
average voyage, 305n
areas fished by south coast schooners, 54-5
August hurricane, 113
'bad fish' and earnings, 91, 115
Bait Intelligence Service, 306-7n
bait provisions: herring, 72-3; caplin, 79; caplin spoilage, 80, 112; squid, 93; scarcity and delays, 93, 307n
Bank Fisherman's (insurance)Fund, 304n
barometer, 175-6
boys hired as lumpers in port 41-2, 52, 53, 309n
bumper trips, 22-3, 232, 234
catch records, 309n, 314n
catch spoilage, 115
changing fishery and community impact, 325-6n
communication
company failure, 126, 169
comparisons: schooner fresh vs. salt fishing, 222-3; Newfoundland vs. Nova Scotia fishing, 104, 111-2,

114-7, 211, 219-21, 307n; communication equipment, 227; engine power, 193-4, 211, 214; Lunenburg vs. Newfoundland skippers, 185; powered and unpowered schooner operations, 170-1, 181, 205; schooner designs, 169, 172
competitive tensions: among dory crew, 95-6
conflict between schooner crews, 181-2
congestion on fishing grounds, 78
contact with foreign fishing vessels, 78
contribution to outport community, 168
cooperation between vessels, 168, 171, 198, 230
cooperation between banks- and Labrador inshore fishers, 143, 148
'counting fish', 10 (see share systems)
crew accomodations, 195, 213
crew management, 103, 284
crew resistance to captain, 104, 117-8
crew turnover, 220, 224, 235

340

dangers:8-9, 68, 110-4, 120, 124; bowsprit 'widowmaker', 88; capsizing, 120; collision, 96-7, 226-7, 231; deck hoisting engine, 88, 91; drowning, 88, 312n; fire, 120 overwork and exhaustion, 90-1; going astray, 86-8; overloaded, 90-1; running aground, 109, 147, 208, 238; sea swept decks, 120, 226; submarines, 80-1, 191, 196-8, 200, 208, 221, 226, 316n; swamping dories in heavy sea, , 85-7, 96; unreliable skipper, 100-1
death reporting, 326n
decline of banks schooner fishing industry, 325n
deck engine, 88, 91, 304n
deploying dories, 86, 199-200
discharging spring trip, 79
displaced by groundfish trawlers, 2
docking for inspection and repair, 80, 141, 153-4, 213, 224, 239
dory fishing gear, 70, 301n
drawing dory berths, 179, 183
'dress crew', 106
earnings 'statement', 91, 199-200, 218-9
earnings uncertainty, 119, 123-6, 130, 149-50, 218-9, 301n, 316-7n
early Fortune Bay dory and crew complement, 48
engaged in Fortune Bay herring trade, 54
end of voyage, 81-2
engine power, 167, 169-70, 224, 227, 312n, 314n
equipage, 72
equipment failure, 99, 108-9, 157-8, 161, 164, 174, 186, 207-8, 213
fall freighting, 50
fair trip, 78, 123
family and occupation, 104, 167-8, 194-5, 199-200, 295n
fishing at anchor, 75-7, 90, 114; underrunniing gear, 94-5
fishing 'under sail', 183-4
fleet decline, 273
fog, 96-7, 226, 232
food provisions, 72, 115-6

foreign banks fishing, 78, 249
foreign-going, 56
foreign-going skippers, 310-1n
'freighting' or 'coasting'(off season work), 97-9, 124-8, 130, 132, 209-10, 212-15, 220, 224, 235
fresh fishing, 101-2, 212, 220-34
good voyage, 93
halibut fishing by American and Nova Scotians, 54
ice conditions, 92, 226
illness among crew:constipation, 164-5; flu, 101; rheumatism, 213-15; typhoid, 87, 97, 112
insurance regulations, 207
'jackboat' schooners, 81, 298n
labour scarcity, 191, 201-3, 208, 211, 326n
Labrador fishery, 312n
lack radio and wireless communication, 58
launching vessels, 180
markets, 89, 220, 234, 241, 304-5n, 309n
mealtime, 116, 172-3
mechanisation displaces sailmaker, 3
menace to navigation, 188
mobility advantages over inshore fishing, 39
naming vessels, 26, 56, 179-80, 308n
navigation skill, 110, 130-40, 161, 177, 232, 310n
Newfoundland built, 56
Nova Scotia built, 26, 102
overwork and exhaustion, 90, 90, 115, 217
owner and captain expectations, 68
owner directs captain to bait source, 93
ownership and investment, 106, 118
owners purchase caribou meat for crews, 29-30
participation in port celebrations, 25
passage home, 83
poor catches and earnings, even hunger, 53
power of owner and captain, 68, 93, 219
preparations for banks voyage, 69-70
radar, 230
radio telephone, 227-8

341

ranking dories and men (high, low), 68, 78, 82, 92, 94-6, 142, 151-2, 207, 217, 305-6n, 308-9n, , 311n, 315-6n
record landings, 199, 224
recreation, 303n
recruitment: kin influence, 52; buying a berth, 64-5; by wire, 65, 88-9, 101, 215; crew origins, 61; criteria, 59, 64; first step (see kedgie), 40; other, 106-7
return to port when crew members die, 58
roles (crew): cook, 70, 77, 97-8, 106, 139, 172, 220, 224, 235; dory crew, 75-7; dory skipper, 85-6, 89; 'dress-crew', 87, 106, 114, 304n; engineer, 204, 207, 226; 'kedgie', 315n; mate ('second hand'), 93-4
Sabbath Day, 323n
safety (radio), 228
sail power: disadvantage, 80
seek bait, 33, 48, 93
seek ice, 48-9
set of sails, 48
settlement (earnings), 71, 82-4, 116, 212, 303-4n
share systems (crew): the count, 10, 62, 68, 71-2, 82, 104, 106, 114-6, 130, 141, 179, 295n, 304n, 311n; learned from French fishers, 71; 'wage' guarantee, 70-1, 82-3, 89; equal shares, 179; risks and tensions created by the count, 94-6, 104, 117-8, 120, 141-2
shipping paper, 71-2, 301-8n
shore processing, 212

skipper authority, 283-4
skipper bookkeeping, 223
skipper competition, 300n
skipper drives men, 90
skipper files (insurance) 'protest', 315n
skipper buys shares in schooner, 106-7, 308n
skipper worries, 109-10, 130, 151-2, 153-5, 174-5
smuggling, 74, 105, 116, 124, 302n
south coast schooner ports, 53-5
St.Pierre, for cheap, duty-free goods, 74, 99-101; petty smuggling, 74
Storm damage, 146
superstitution, 180, 184-6
taffrail log, 136
tallyboard, 311n
torches, 232
tricks ('stunts'), 68, 71, 212, 228-9; extra lines, 94-5
vessel insurance, 189-90, 207
vessel meanings, 180-1
waste during on board processing, 90
watch keeping, 73-5, 160
water provisions, 73, 99-100
weather delays fishing, 93
winter supplies at settlement, 83
women cure schooner fish ashore for extra income, 30, 168, 190
women's and home responsibilities, 104
woodcutting, 126
work values, 67, 84, 95-6, 108, 130, 143, 302n

Deep sea fishing II: 'Dragger' or trawler fishing (241-79)

accidents, 270-2, 321n, 324n
accomodations, 244, 247, 249, 263
barometer, 254, 276
boat design, 248, 292
boat construction in Newfoundland, 325n
bottom 'snags', 256-8, 261, 267-8

bumper catches, 257-8, 260-2, 279, 324n
catch disgard ratios, 320n
catch performance comparisons, Appendix, 330-3n
catch records, 318-9n, 323n
catch species variety, 319n

community support in crisis, 291-2
comparison with banks schooner fishing, 242-3, 249, 255
competition between skippers, 264-5, 285
conflict with inshore fishery, 323n
conflict with banks schooner fishery, 318n, 325n
conservation need, 325n
cooperation between trawler skippers, 249
co-venturer myth, 243, 291
crewing, 245, 264
culling species, 320n
dangers, 244, 275-6
direction finder, 270, 275
discharge equipment ashore, 252
docking, 261-2
draggers enter Newfoundland, 2, 316n, 318n
ecological range, 242
earnings, 244, 251, 253, 262-3, 319-20n, 321n, 324n
emergency, 269
employment alternatives, 292
engine power, 244, 247-8, 262
family and occupation, 245, 258-60, 275, 284, 288-91
finding fish, 320n
fishermen knowledge, 242-3, 267-9, 278
fish plants, 248, 256
foreign schooners, 274
foreign trawlers, 249, 272-3, 278, 324-5n
gear (trawl) deployment, 319n
gear protection, 322n
impact on inshore fishery, 273
inexperience, 245-7, 249, 251
insurance for crew, 321n
integration with shore processing, 244
International Commission for Northwest Atlantic Fisheries (ICNAF), 325n
job scarcity, 243, 273

Loran, 275
luck, 287
marine disaster relief, 291
markets, 261, 279
media interest, 242, 316n, 323-4n
merchant-skipper relation, 244
mishaps, 263-4
naming boats, 260, 317-8n, 326n
navigation, 248, 320n
overwork and exhaustion, 322n
owner priorities, 270, 275-6
organizational persistence, 243
processing plant capacity, 323n
radio communication, 244, 249-51, 254, 257-9, 264-5, 269-72, 275, 316n
recruitment, 245, 248
redfish, 256-7, 278
refit, 260, 279
retirement, 292
roles:cook, 245-8; engineer, 253
Sabbath Day, 323n
seaworthiness, 253-4
saltfishing, 265-6
skipper dominance, 243
skipper drives men, 321-2n
skipper as gatekeeper-patron, 285
skipper worries, 249, 255
stock destruction, 243, 253, 255, 257, 264-5, 272-3, 278-9, 320n, 325n
target species, 249, 257, 264-7
tragedy, 275-7, 291-2
transition from schooner fresh fishing, 212, 224
trawler lost at sea, 326n
trawler turn-around time, 252
tricks, 264-5
unionization, 296n
values, 284
watch, 252
'weighback' (undersize fish & ice), 273
winch operation, 271
winter fishing, 253
work regime, 243, 245, 252, 262-3, 319n, 321n.

Department of Natural Resources, Newfoundland, 162
Department of Public Works, Canada, 292-3
Department of Transport, Canada, 292
Doctors: scarcity, 20-1, 62, 83-4, 111-2
Depression (era), 119, 12530, 150, 310-1n
Dole, 16, 27, 308n
Dreams, 24
Drum fish, 208
Earthquake(1929), 121-2
Education, 19, 20, 36-7, 43, 290
Emergency Radio Alert, 162-3
Emigration, 310n, 317n
Engine power, 43, 53, 167, 183, 188, 192, 200-4, 212, 238, 316n
Fairbanks Morse, 224
Faith, 302n
Fisheries: frontier resources, 35; global trends, 11; inshore vs. offshore and relations, 38-9, 148-9; intensification, 192, 211-2, 242-3, 253; offshore expansion, 192; stock destruction and human tragedy, 11; lobster, 28; saltfish trade, 35
Fishermen (fishers), deep sea: attitudes towards 'count' share system, 68, 115, 141; inter-generational, schooner-dory vs. trawler, comparison, 5; attraction of foreign trade, 88; captaincy requisites, 99, 119, 131, 143-4, 159; captains become merchants, 35; captains take owner directions, 93; coastal origins, 17, 53, 188; communication with home via telegram, 77; credit system dictates conformity, 68; crew rebellion against skipper, 100-101; domination by vessel operators, 10, 31, 68, 219; first trial, 68; formal education, 67; injuries (infected fingers), 70; inshore vs. offshore and polarization, 38-9; kin ties and skipper recommendations, 99, 101, 149; knowledge and skills, 36, 67, 85-6; literacy, 296n; memorial to those lost, 26; 'laddio' skipper, 99-101; owner and skipper expectations (effectiveness, obedience, reliability, sobriety), 68, 71, 143-4; merchants compete for good captains, 98; political voice, 10, 219; practical-moral indoctrination in skills and relationships, 36; recreation in port, 69, 77-9, 87, 96, 106, 221-2; relations between Catholics and Protestants, 79; religious tolerance; reputation, 55, 67, 71, 85 ('highliner'), 94, 98, 167, 199, 230, 232, 305-6n, 315-6n, ; success requisites, 67-8, 94-5; values, 36-7; self-reliance, 68
Fishery Board of Scotland, 307n
Fishing grounds: 'Bethel Shoal', 257-8, 268-9, 323n; Bonne Bay, 54; French Shore, 54; Cape George, 54, 120; Cape St. Mary's, 54, 147; Grand Banks, 42-247 *passim*; Gulf of St.Lawrence, 108, 111, 197, 204, 267; Hermitage Bay, 266-7; 'Hump', 229; 'Jessie Ryder's Shoal', 323n; Labrador, 125, 143, 147-8, 152, 181-2, 217; Rose Blanche Bank (also Western Shore) 30, 45, 57, 182, 197, 200, 203-8, 220; 'Round Hill' (Island), 181; Sable Island, 97; St. Pierre Bank, 2, 81, 276, 320n; Southern Labrador, 42-3, (Forteau Bay, Greenly Island, West St. Modeste, Blanc Sablon, to Mecatina Island) 54, 55; Virgin Rocks, 268; Western Banks ('Quero or Bankquero, Missaine or Mizzen, and George's)55, 78, 96, 99, 109, 111, 113, 117, 182, 195-7, 213; 'Western Shore' (see Rose Blanche Bank)
Fall trips: 55, 80
Fisheries and Oceans, Canada, 318n
Flounder, 261
Fort Pepperell, 271
'French Shore', 300n
French three mast vessels at St.Pierre, 74; on Bank Quero, 78
Gender roles: 30-1, 36, 40-1, 47

German submarines, 200, 303n, 309n
Grenfell Foundation bursary, 305n
'Growth centres', 241
Haddock, 249, 253, 261, 278
Halibut, 54, 269
Homecoming, 24-5, 83, 85
Hospital, 111
Humour, 56
Hunting, 16, 26, 28-9, 50, 53
Hurricane damage, 21
Ice-making and storage, 8-9;
Illnesses: appendicitis, 45; blood poisoning, 20; constipation, 164; epidemic; diptheria, 62-4, 84; seasickness 40, 73-8, 82, 84; shellfish poisoning, 58; smallpox, 21; Spanish influenza, 83-5; typhoid, 87, 96-7
Inshore fishery: seasonality, 30, 32; changing technology and organization, 10; cod (gill) nets, 30; handlining, 30; herring gill nets and seines, 33-4; heritage, identity and values, 11; impact of banks schooner fishing impact on inshore lifestyle, 39; vs. trawler, as extreme industrial labour regime, 10; trawls, 30; work regime comparisons, 10; other, 241, 298-9n, 322-3n, International Commission for Northwest Atlantic Fisheries, 243
'Kedgie', 40, 45, 51-2
Industrialism and fish stock destruction, 191, 211, 241-2
Labrador coastal fishery, 311-2n
Labrador fish quality market impact, 89
Life story: representativeness, 9-10; lessons, 9; inter-generational understanding, 10
Literacy: 19, 25, 296n, 310n
Lobster trade, WWI, 49
Local trader, 44, 298-9
Long Harbour, Fortune Bay: 15, 17, to caribou hunting ground and salmon rivers, 17-8, 50, 59; berry picking, 32; cod fishing, 57; freezes over 18, 29, 50; spring fishery after thaw, 61 woodcutting, 33

Lunenburg Marine Railway, 315n
Lunenburg Marine Insurance Company, 315n
Mail service, 44
Maritime Museum, Mystic, Conn., 311n
Maritime shipping, 129
Medical treatments: 'brimstone', 20; Castor Oil, 164; sulphur and molasses, 297n; vaccination, 21
Memory: emotive content, 8-9; limitations, 8, 319n
Mi'kmaq, 299n
Missionary educators: Anglican, 19; Congregational, 19
Mount Allison University, 245, 279
Mourning, 22, 24, 26-7, 58, 92
Narrows, St.John's, 221-3
Neighbours: mutual help, 50
Newfoundland College of Fisheries, 311n
Newfoundland Federation of Fishermen, 326n
Newfoundland Fisheries Development Authority, 293
Newfoundland Outport Nurses Industrial Association (NONIA), 306n
Newfoundland Supreme Court, 313n
North Atlantic Fisheries Convention (1949), 325n
Occupations: blacksmith, 3; carpenter, 44, 206; domestic ('in service'), 38; game warden, 43; 'kedgie, 40, 45, 51-2; lumper, 41-2; postal, 44; sailmaker, 3; small trader 44; teacher, 83-4; telegrapher, 61, 83
Orange Society, 26, 92
Play, 22
Portuguese, 270
Prince Edward Island: trade with Anderson's Cove, 16; for produce, 97-8, 105
Prince of Wales Collegiate, 279
Prosperous fishing times, 35, 58
Recruitment to British navy, 48
Regatta, St.John's, 260
Religious Denominations: Anglican, 19; Catholic, 79; Congregational,

20, 296n; Methodist, 79; Salvation Army, 69, 79, 85
Resettlement, 20
Salmon fishery, 44
Salvation Army, Grand Bank, 69
Sealers, 131
Foreign saltfish trade, 23, 28-9; Danish carrier, 73; Newfoundland three mast vessels, 78
Scientists, 93, 242-3, 260, 278, 318n
Settling accounts, 31
'Shipping out', 27-8
Spanish 'pair trawlers', 328n
Spring trip (fishery), 46, 48, 53, 61, 170
Social anthropology: research method 5
Spirituality, 38, 168, 172-3
Starvation, 310n
Suicide, 24
'Swifton's Days', 315n
Terms of address, 21
Tidal Wave (1929), 309n
Tradition, 37-8
"Traditional evironmental knowledge", 37
'Times', 22

Tragedies, 21-254 *passim*
United States Naval Dock, 324n
Values: community, 36-7; family, 27, 36-7, 59; frugality, 30-1; hard work, 9, 37, 64, 76; honesty, 9; initiative ('eagerness'), 19, 20, 30-1, 36, 40, 60; reliability, 61; religious faith, 38; self-reliance, 27, 37; self-sufficiency, 28; sharing, 37, 41, 59
Vichy France, 192
Walsh Commission, 311n
War prize vessel, 192
Wedding celebration, 308n
West Indies, 208, 210
'West Indies prices, '91, 115
Winter (banks) fishing, 25, 45, 51, 57, 69, 89-90, 92, 94, 209
Winter mooring, 106, 113-4, 118-9, 126-8, 150, 154
"Workmen's Compensation Act", 291
World War I, 39, 48-9, 50, 56, 58, 60-1; ends, 84; submarine threat, 79, 303n; Nova Scotian vessels sunk, 80-1
World War II, 168, 187, 191, 208, 211-2, 220-1, 226-7, 234, 310n

GLOSSARY

Baiting(s): The salt fish banks fishing year or 'voyage' had three phases, 'trips' or 'baitings,' each based upon a different baitfish species, namely herring (winter-spring), caplin (summer), and squid (late summer-early fall). But a schooner usually restocked its supply of the same bait species several times during each species or trip phase before that 'baiting' ended.

Banker: Short for banks schooner.

Banking (or banks) schooner: "A vessel engaged in cod-fishing on the Newfoundland offshore grounds" (Story,Kirwin, & Widdowson 1990, p.21).

Banks fishery: Cod fishery conducted on Newfoundland offshore grounds.

Barrack head: The partially sheltered forepeak deck area of the schooner ahead of its gangway and windlass or donkey engine.

Bulked: Piled or stacked, as with caribou carcasses.

Bumper trip: A very large catch

Caplin trip: The second major phase in the usual fishing year or 'voyage' in the Fortune Bay banks schooner and dory fishery. This bait phase is based upon iced caplin and lasts about three to four weeks, roughly from early to mid June and August.

Carpot: Large wood framed box for holding captured live lobsters near-shore until processed or shipped.

Coasting: Sailing between outport settlements and other coastal points usually for trading purposes. Also 'freighting'.

Coasting crew: Seamen shipped aboard a vessel usually engaged in coastal trade. Newfoundland trading vessels, and vessels in the foreign trade, shipped about five or six hands (cf.Story,Kirwin, & Widdowson 1990, p.103).

Co-adventurer: Also co-venturer. The system or relationship between vessel owner and fishermen where both share responsibilities, risks and returns of fishing effort. Newfoundland's banks schooner fishing merchants and vessel owners controlled ownership, terms and ultimate conduct of operations, value of fish landed, and market access. Meanwhile, it seemed crews and solitary individual dory fishers shouldered the burden of misfortune, poor catches, and poor prices on meager rations and often wondered about the equity of this relationship and settlement outcomes.

Coasting crew: Seamen who sail a vessel between ports, usually with cargo, and perform maintenance functions.

Count: Reward or incentive system where each dory fisherman's earnings were based upon one half the value of the number of fish landed aboard the mother schooner by his dory alone. The vessel owner determined the ultimate valuation after the fish were landed, processed, and marketed.

Crack out day: When captive live lobsters are cooked and canned.
Dragger: A vessel that tows a cone shaped net or trawl. The vessel tows two cables, each one attached separately to an angled otter bord or vane, which, in turn, is attached by cable to each side of the net. This holds the net mouth open. Floats attached to the net's top- or headline provide vertical lift, and a combination of heavy rubber and/or steel rollers (bobbins) on its bottom- or foot-rope enable the trawl to be hauled or dragged along the bottom. In Newfoundland parlance a dragger and trawler are the same.
Dragging: The action of towing the heavy trawl along the sea bottom to catch fish.
Drive works: To skylark.
Dress: To cut, trim and process fish.
Dress crew: Aboard Nova Scotian vessels in the banks cod fishery described here, it included a gutter, header, splitter, idler, and salter.
Drive with the sea(s): Run a vessel in the direction of wind and sea.
Dummy schooner: A schooner powered only by sail.
Fish under sail: In banks dory fishing, the practice of dropping off and retrieving dory crews and their catch from the unanchored schooner mothership.
Flake: Frame for drying cod fish.
French shore: The area from Cape St. John to Cape Ray, Newfoundland (see Story, Kirwin & Widdowson 1990:202).
Frozen baiting: The frozen herring used during winter-spring banks cod fishery.
Groundfish trawler: Vessels equipped to fish bottom dwelling (vs. pelagic) fish species with towed catching gear. See 'dragger.'
Gutter: Informal role when dressing cod fish, he splits the fish from anus to head and removes internal organs.
Handline: Weighted line, usually with baited hook, fished by hand overside, whether from dory, schooner, or other vessel.
Head the sea(s): Bring or hold a vessel's bow into the oncoming wind and sea.
Header: Informal role in 'dress crew' when dressing cod fish. The header severs the head (and may remove entrails).
High and low: As in 'high dory' and 'low dory', rank based upon catch performance when banks fishing on the 'count' share system.
Highliner: When banks dory fishing, it is the dory crew that lands most fish. More generally, it is the individual who lands the most fish among competitors.
Idler: Informal crew role in 'dress crew' when processing fish. After fish are gutted, headed, and split during dressing, the idler washes the fish before it is put into the vessel's hold and salted.
In bulk: In reference to dressed cod salted and stacked in fish stores, before transfer to barrels or vessel holds for shipment.

Jackboat: Sloop rigged, cod fishing vessel usually with one to four dories and gear to fish shore grounds.

Jibe over: In sailing, the vessel is brought on a new course by rudder correction and tacking, shifting the sail boom(s) and jib from one side of the vessel to the other.

Jiver: The track and block assembly astern into which the main sheet is run for control of the main sail.

Kedgie: Deckhand, cook's assistant, and sometimes replacement for adult dory fishermen.

Knockabout: Fore and aft rigged vessel having no bowsprit. Its single jib is bent to a stay from the stem head.

Laddio: A carouser, often a drinker.

Light sail: A schooner's topmast sails (gaff topsail, topmast staysail, foregaff staysail, and jib baloon.

Log-loaded: A heavily burdened vessel that lies low in the water.

Longers: Long sticks of wood used to build structures, as in frames or flakes for drying fish.

On account: In a credit relationship with a fish merchant/vessel owner, or other merchant.

Over to: A change of course, especially in sailing, when the sail position is shifted when turning from starboard to port, or vice versa.

Pin maul: A wood handled maul with two iron working ends, one pointed (often used to drive holes in barrels) and the other blunt (for use in driving wedges and other things).

Quintal: A standard measure of weight in the cod fishery. One quintal equals 112 lbs. of washed and dried salt fish.

Reserved gear: Equipment designated and held for a specific vessel.

Run: In the banks cod fishery it is the complete cycle of dories setting baited trawl, return to mothership for grub or mugup before hauling trawl back and removing catch, rebaiting and relaying the trawl (when 'underrunning'), then discharge aboard vessel.

Salter: Informal crew role when storing dressed cod. The salter applies salt to dressed fish prior to its storage below deck.

Second hand: Mate.

Sed: About one fathom of light line attached at one end to main trawl line and to hook at other; also 'sud'.

Settle-up time: When individual fishermen meet with fish merchant employers to determine their earnings (or indebtedness), usually upon completion of the fishing year or voyage.

Share(s): Generally, it is any one of many schemes (including the 'count') designed to reward fishing effort. In the banks cod fishery, it is an average of a crew's portion of the value of their vessel's, or their individual dory's, fish landings after subtraction of individual expenses and the crew part of operating expenses (e.g. bait, ice, food).

Ship out: Here, to join a vessel departing home or port for the banks or other destination.

Side trawler: A vessel that deploys, tows and retrieves its otter trawl groundfish gear, usually from its starboard side.

Skeleton refit: Speaking of trawlers, it is a brief and less than complete reconditioning of a working vessel.

Sound(s): Cod air bladder.

Soundbone: Cod backbone.

Splitter: Informal crew role when dressing fish. The splitter removes the sound or back bone and splits and flattens the cod prior to washing and salting.

Spring trip: It is based upon frozen herring and the first phase in the usual fishing year or 'voyage' in the Newfoundland banks schooner and dory fishery.

Stern trawler: A vessel equipped with a stern chute for deploying and retrieving groundfish and/or midwater trawl fishing gear. During the 1960s many European groundfish vessels operating in Northwest Atlantic waters were also 'factory' stern trawlers, i.e. equipped with fish filleting machines and freezing equipment typical in fish processing plants ashore.

Summer trip: See 'caplin trip.'

Swing the four lowers: In sailing, under fair winds, the main, fores'l, and jib and jumbo are up and out, driving the vessel before the wind.

Taffrail log: A device towed to measure distance covered over water.

Tally board: Any board used to record the count of fish landed on board by individual dory crews working from banks vessel.

Time(s): Festive social occasions accompanied by some combination of feasting, music, singing, story telling, and dance, e.g. to celebrate a wedding or other happy event.

Trawl: The term is used in several fisheries for quite different technologies. In banks schooner and dory fishing, it is the buoyed line with many spaced short lines and their hooks deployed by dory crews. In trawler or dragger fishing, it is the large (otter) net and related gear towed by the vessel over the bottom.

Trip(s): One or more baitings; usually the time between taking on bait and landing a catch for processing. In the banks schooner fishery, a trip is tied to a particular bait (herring, caplin, squid). Hence a vessel may fish and land several baitings (under herring, caplin, or squid) before that trip is completed.

Under-run the trawl: When fishing from anchored mother vessel, each banks dory crew anchored and set its trawl, later drew the trawl line over their dory bulwarks, removed the catch and rebaited the line, then let it sink again to continue fishing. The dory then returned to the mothership to discharge the catch before returning to haul the trawl again. This may continue for several days if the location is productive.

Weather edge: As in banks dory fishing, the windward end of anchored trawl gear.

Western shore: In this volume, it is the area from Fortune Bay west to Cape Ray.

West Indies price: The lowest market price for salt cod.

Winter fishery: In banks cod fishery it was usually based upon frozen herring, ran from as early as January to April and carried out off Newfoundland's southwest coast or 'Western shore'.

Winter trip: Also see 'spring trip'. First major fishing phase in Fortune Bay banks schooner fishing year or 'voyage.' Based upon frozen herring. It usually began by April 1st and ran to the summer (by June) bait phase.

Voyage: The fishing year. In the south coast Newfoundland banks fishery it usually began with the 'spring trip' (about April 1st), followed by the 'summer trip' on iced caplin, then concluded with the 'fall trip' on squid bait. This baiting sequence brought the vessel off the southwest coast and Western banks, eastward to the Grand Banks, then to the northeast Labrador coast. The voyage ended in October-November at home port where crews 'settled up' with their merchant patrons and "co-venturers".